Computer Holography

Acceleration Algorithms and Hardware Implementations

Computer Holography

Acceleration Algorithms and Hardware Implementations

Tomoyoshi Shimobaba
Chiba University, Japan

Tomoyoshi Ito
Chiba University, Japan

CRC Press is an imprint of the
Taylor & Francis Group, an **informa** business

CRC Press
Taylor & Francis Group
6000 Broken Sound Parkway NW, Suite 300
Boca Raton, FL 33487-2742

First issued in paperback 2023

ISBN 13: 978-1-032-65225-2 (pbk)
ISBN 13: 978-1-4822-4049-8 (hbk)
ISBN 13: 978-0-429-42800-5 (ebk)

DOI: 10.1201/9780429428005

Publisher's Note
The publisher has gone to great lengths to ensure the quality of this reprint but points out that some imperfections in the original copies may be apparent.

Library of Congress Cataloging-in-Publication Data

Names: Shimobaba, Tomoyoshi, author. | Ito, Tomoyoshi, author.
Title: Computer holography : acceleration algorithms & hardware / Tomoyoshi Shimobaba and Tomoyoshi Ito.
Description: Boca Raton, FL : CRC Press/Taylor & Francis Group, 2018. | Includes bibliographical references and index.
Identifiers: LCCN 2018029931| ISBN 9781482240498 (hardback : acid-free paper) | ISBN 9780429428005 (ebook)
Subjects: LCSH: Holography--Data processing. | Holography--Mathematics. | Holography--Equipment and supplies.
Classification: LCC QC449 .S55 2018 | DDC 621.36/750285--dc23
LC record available at https://lccn.loc.gov/2018029931

Visit the Taylor & Francis Web site at
http://www.taylorandfrancis.com

and the CRC Press Web site at
http://www.crcpress.com

Contents

List of Figures

List of Tables

Preface

Holography was invented in 1947 by Hungarian physicist Gabor. Holography can record three-dimensional (3D) information of light on a two-dimensional recording medium (hologram) by using light interference and diffraction. If the hologram is irradiated with light, the recorded 3D image is faithfully reconstructed. With this characteristic, holography is said to be the ultimate 3D imaging technology.

The theory of holography was studied and almost completed by the 1970's. Meanwhile, since the 1970's, the fusion of holography and computers (computer holography) has begun. With the development of computers, the applications of computer holography have expanded. Currently, the following research is vigorously progressing: holographic display, digital holography, computer-generated hologram, holographic memory and optical encryption.

In these applications, real-time processing is required and the amount of information has dramatically increased with the development of computers; therefore, the acceleration of the calculation is one of the important themes.

This book aims at the acceleration of computer holography. Computer holography is based on diffraction calculation, so the key point is how fast the diffraction calculation is processed. Various algorithms have been devised for this acceleration. In this book, we introduce acceleration algorithms.

In addition, the unique point of this book is to explain hardware acceleration. Field programmable gate arrays (FPGAs) and graphics processing units (GPUs) are promising hardware. We explain how to implement calculations for computer holography on these hardware components.

The organization of this book is as follows. Chapter 1 describes the scalar wave necessary for computer holography and the principle of holography. In Chapter 2 we explain the principle of diffraction calculations which are often used in computer holography and the numerical implementation of diffraction calculation using C language. In computer holography, MATLAB® is frequently used as a programming language, but this book uses C language because C language is suitable for the hardware implementation.

In Chapter 3, we describe acceleration algorithms used in holographic displays. Chapter 4 describes digital holography. In digital holography, 3D information of existing objects is recorded as hologram data in a computer using an image sensor. The 3D information can be reconstructed in a computer by diffraction calculation from a hologram with a computer. The application to microscopes is actively performed.

Chapter 5 introduces various applications of computer holography. In this chapter, we will first explain a phase retrieval algorithm that is used for finding vanishing phase information and optimizing holograms. Next, we will explain holographic memory, projection and the applications of deep learning to computer holography. Deep learning has been studied vigorously in recent years, and its applications have begun in computer holography. We also expect deep learning to be one of the core technologies in computer holography. Chapter 6 describes the hardware implementations of computer holography. We will explain how to implement hologram calculation and diffraction calculation in FPGA and GPU.

This book is based on our lecture at Warsaw University of Technology and our review paper (IEEE T. Ind. Inform. **12**, 1611–1622 (2016)) on computer holography. We hope this book will help students and researchers who study computer holography from now on.

We gratefully acknowledge our co-workers: Prof. Takashi Kakue, Prof. Naoki Takada, Prof. Nobuyuki Masuda, Prof. Minoru Oikawa, Dr. Yasuyuki Ichihashi, Prof. Takashige Sugie, Prof. Atsushi Shiraki, Prof. Yutaka Endo, Dr. Ryuji Hirayama, Dr. Hirotaka Nakayama, Prof. Takashi Nishitsuji, Prof. Yuki Nagahama and Prof. Naoto Hoshikawa.

We gratefully acknowledge our co-worker, Prof. Michał Makowski (Warsaw University of Technology), who gave us the opportunity to give the lecture at his university.

In addition, we gratefully acknowledge Prof. P. W. M. Tsang (City University of Hong Kong) and Prof. T. -C. Poon (Virginia Tech) who gave us the opportunity to write the review paper based on this book.

Tomoyoshi Shimobaba is an associate professor at Chiba University, Japan.
Tomoyoshi Ito is a professor at Chiba University, Japan.

Author Biographies

Tomoyoshi Shimobaba is an associate professor at Chiba University. He received his PhD degree in fast calculation of holography using special-purpose computers from Chiba University in 2002. His current research interests include computer holography and its applications. He is a member of SPIE, OSA, the Institute of Electronics, Information, and Communication Engineers (IEICE), OSJ, and ITE.

Tomoyoshi Ito is a professor at Chiba University. He received his PhD degree in special-purpose computers for many-body systems in astrophysics and in molecular dynamics from the University of Tokyo in 1994. His current research interests include high-performance computing and its applications for electroholography. He is a member of ACM, OSA, OSJ, ITE, IEICE, IPSJ, and the Astronomical Society of Japan (ASJ).

1 Basics of holography

Computer holography needs to treat light propagation on a computer. Light propagation should be addressed as vector waves based on Maxwell's equations; however, the treatment of vector waves is complicated and requires much computation time and memory. Fortunately, in most applications of computational holography, it can be treated as a scalar wave, making the calculation simpler. This chapter describes the basics of scalar waves and holography.[i]

1.1 SCALAR WAVE

A light is a kind of electromagnetic wave. Electric and magnetic fields are transverse waves, and propagate in a direction perpendicular to the vibration direction. Strictly, a light should be handled based on Maxwell's equation. The electric and the magnetic fields are expressed as vectors, and are called **vector waves**. It is generally cumbersome to handle vector waves because it is necessary to calculate all the vector components.

When the size of objects to be irradiated by light is larger than the wavelength of light, the light can be handled approximately as a **scalar wave** since each vector component behaves similarly. That is, only one component needs to be handled. In the following, light waves are handled as scalar waves.

Consider an expression of scalar wave propagation in a three-dimensional space, starting from wave propagation in a one-dimensional space. When the one-dimensional wave $u(x,t)$ where x is the coordinate and t is the time propagates to the right direction with the speed v as shown in Figure 1.1, the wave is expressed as

$$u(x,t) = f(x - vt). \tag{1.1}$$

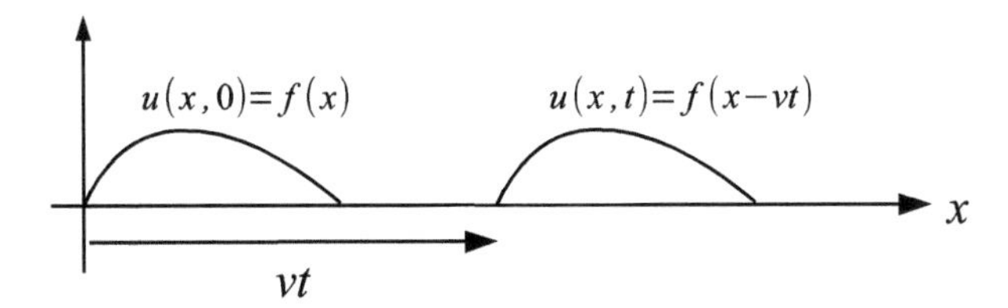

FIGURE 1.1 Wave propagation in a one-dimensional space.

Although the function $f(\cdot)$ representing the shape of the wave can take any

[i]If we want to study the contents of this chapter more deeply, we recommend a famous textbook by Goodman [13].

functions, light waves are generally expressed by sinusoidal waves as follows:

$$\begin{aligned} u(x,t) &= a\cos(kx - \omega t + \phi), \\ &= a\ \mathrm{Re}\{\exp(i(kx - \omega t + \phi))\}, \end{aligned} \tag{1.2}$$

where a, k, ω, and ϕ are the amplitude, **wave number** ($= 2\pi/\lambda$ where λ is the wavelength of the light), angular frequency and initial phase, respectively. In the second equation, we can readily handle the wave by using complex numbers. For this, we use **Euler's formula** as follows:

$$\exp(ix) = \cos(x) + i\sin(x). \tag{1.3}$$

$\mathrm{Re}\{\cdot\}$ means to take the real part of the complex value. This expression can be further separated into time and space variables as

$$\begin{aligned} u(x,t) &= a\exp(i(kx - \omega t + \phi)), \\ &= A(x)\exp(-i\omega t), \end{aligned} \tag{1.4}$$

where $A(x)$ is defined as

$$A(x) = a\exp(i(kx + \phi)). \tag{1.5}$$

This is called **complex amplitude**. As we will see later, in computer holography, terms related to time are not important and it is enough to handle only the complex amplitude.

Wave propagation in two- and three-dimensional space can be easily derived from the one-dimensional wave propagation. Let us consider the two-dimensional wave propagation.

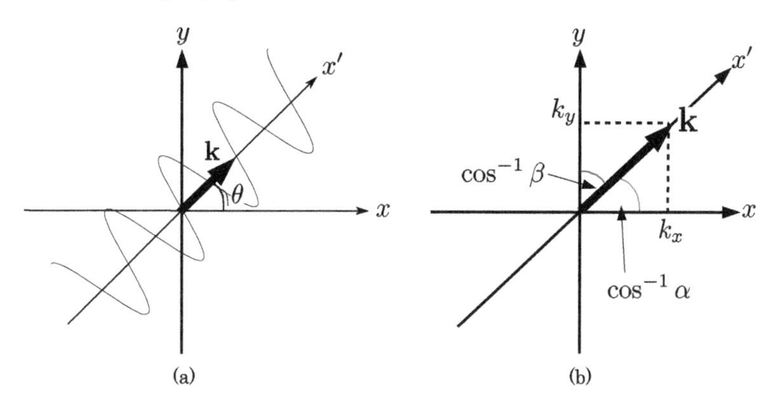

FIGURE 1.2 Wave propagation in two-dimensional space. Planar wave propagates at the angle θ. (b) Planar wave expressed by direction cosine.

For example, as shown in Figure 1.2, a scalar wave propagating along the x' axis tilted by θ with respect to the x-axis can be considered as a one-dimensional wave propagation on this axis. Therefore, we can express it as

$$u(x',t) = a\exp(i(kx' - \omega t + \phi)). \tag{1.6}$$

When observing $u(x', t)$ from x-y coordinates, the two-dimensional wave propagation can be expressed as

$$\begin{aligned} u(x, y, t) &= a \exp(i(kx \cos\theta + ky \sin\theta - \omega t + \phi) \\ &= a \exp(i(k_x x + k_y y - \omega t + \phi)) \\ &= a \exp(i(\mathbf{kx} - \omega t + \phi) \end{aligned} \quad (1.7)$$

where $\mathbf{x} = (x, y)$ represents the position vector, and $\mathbf{k}$ is the **wave vector**. The wave vector is a parameter that determines the propagation direction of the wave. The wave vector $\mathbf{k}$ is defined as

$$\mathbf{k} = (k_x, k_y) = (k \cos\theta, k \sin\theta). \quad (1.8)$$

There is another way of representing a wave vector using **direction cosine**. If we define the angle between the x' axis and the x-, y-axes as $\cos^{-1}\alpha$ and $\cos^{-1}\beta$, the wave vector can be written as

$$\mathbf{k} = (k_x, k_y) = k(\alpha, \beta) \quad (1.9)$$

where α and β are called the direction cosines.

By using the position vector and the wave vector, three-dimensional wave propagation can be similarly described as shown in Figure 1.3. The three-dimensional wave propagation is expressed as

$$\begin{aligned} u(\mathbf{x}, t) &= u(x, y, z, t) = a \exp(i(\mathbf{k} \cdot \mathbf{x} - \omega t + \phi)), \\ &= a \exp(i(k_x x + k_y y + k_z z - \omega t + \phi)), \end{aligned} \quad (1.10)$$

where $\mathbf{k} = (k_x, k_y, k_z)$ and $\mathbf{x} = (x, y, z)$. The complex amplitude $A(\mathbf{x})$ is written as

$$\begin{aligned} A(\mathbf{x}) &= a \exp(i(\mathbf{k} \cdot \mathbf{x} + \phi)), \\ &= a \exp(i(k_x x + k_y y + k_z z + \phi)). \end{aligned} \quad (1.11)$$

If we write the angles between the wave vector and the x, y, z axes as $\cos^{-1}\alpha$, $\cos^{-1}\beta$ and $\cos^{-1}\gamma$, respectively, the wave vector using the direction cosine is expressed as

$$\mathbf{k} = (k_x, k_y, k_z) = k(\alpha, \beta, \gamma). \quad (1.12)$$

From the relation of $|\mathbf{k}| = k(= 2\pi/\lambda)$, (k_x, k_y, k_z) are not independent of each other, and have a nonlinear relationship as follows:

$$k_z = \sqrt{1 - k_x^2 - k_y^2}. \quad (1.13)$$

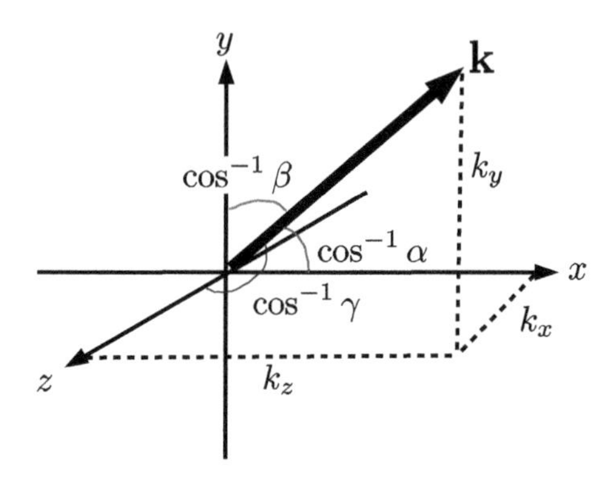

FIGURE 1.3 Wave propagation in three-dimensional space.

1.1.1 LIGHT INTENSITY AND INTERFERENCE

The intensity of light can be detected using photographic films, photodiodes, and imaging sensors such as CCD cameras. The frequency of light is $f = c/\lambda$ where c is the speed of light (about 3×10^8 m/s in vacuum) and λ is the wavelength. For example, in the wavelength of visible light (400 nm $\sim$ 700 nm), the frequency of light in vacuum is about 10^{15} Hz. Unfortunately, there are no imaging sensors that can directly capture light that vibrates 10^{15} times per second, so imaging sensors can only capture the time average of the vibration. This is called **light intensity**. The light intensity of Eq. (1.10) is expressed as

$$I = |u(\mathbf{x}, t)|^2 = |a|^2. \tag{1.14}$$

The holography described in Section 1.2 uses the **interference** of light waves. Let us consider the interference of two waves, $u_1(\mathbf{x}, t)$ and $u_2(\mathbf{x}, t)$. These waves can be expressed as

$$\begin{aligned} u_1(\mathbf{x}, t) &= a_1 \exp(i(\mathbf{k_1} \cdot \mathbf{r} - \omega t + \phi_1)), \\ u_2(\mathbf{x}, t) &= a_2 \exp(i(\mathbf{k_2} \cdot \mathbf{r} - \omega t + \phi_2)). \end{aligned} \tag{1.15}$$

Since the interference is simply the superposition of waves, it can be written as

$$u(\mathbf{x}, t) = u_1(\mathbf{x}, t) + u_2(\mathbf{x}, t). \tag{1.16}$$

When observing this interference with an imaging sensor, we can only obtain the time average of the interference (light intensity). Therefore, we can write it as

$$\begin{aligned} I(\mathbf{x}, t) &= |u(\mathbf{x}, t)|^2 = u(\mathbf{x}, t)u^*(\mathbf{x}, t) \\ &= |a_1|^2 + |a_2|^2 + 2a_1a_2 \cos((\mathbf{k_1} \cdot \mathbf{r} + \phi_1 - \mathbf{k_2} \cdot \mathbf{r} - \phi_2). \end{aligned} \tag{1.17}$$

The time term of the light wave can be omitted. In this book, we do not handle the time term.

1.1.2 PLANE WAVE AND SPHERICAL WAVE

Types of light waves used in computer holography are mainly the spherical wave and plane wave. As shown in Figure 1.4, the light wave emitted from a **point light source** is isotropically spread and the wavefront of the light wave becomes spherical when connecting the parts with equal phases (**equiphase surfaces**). This light wave is called a **spherical wave**. A spherical wave is simply expressed as

$$u(r) = \frac{a}{r} \exp(i(kr + \phi)) \tag{1.18}$$

where r is the distance between the point light source at (x_p, y_p, z_p) and arbitrary point at (x,y,z) in three-dimensional space. r is expressed as

$$r = \sqrt{(x - x_p)^2 + (y - y_p)^2 + (z - z_p)^2}. \tag{1.19}$$

The amplitude of the spherical wave attenuates with $1/r$.

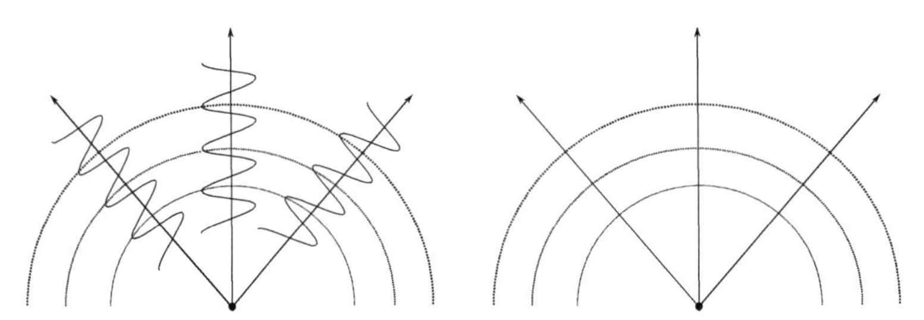

FIGURE 1.4 Spherical wave.

As shown in Figure 1.5, when the wavefront of a light becomes a plane when the parts are connected with equal phases (equiphase surface), this light wave is called a **plane wave**. A spherical wave when a point light source is at

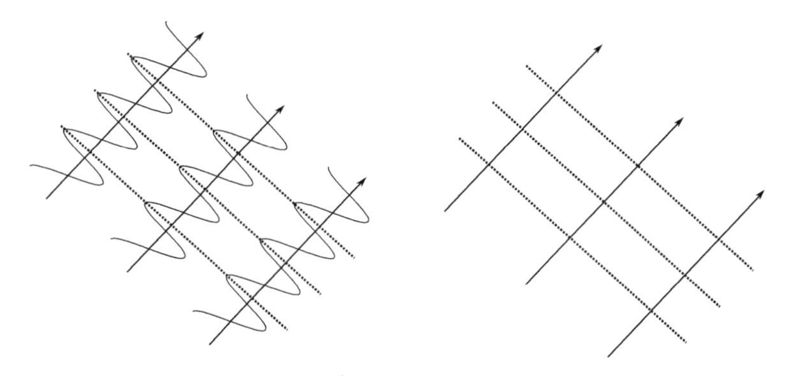

FIGURE 1.5 Plane wave.

infinity can be regarded as a plane wave.

The propagation direction of a light wave is determined by the phase part of the light wave, which is generally expressed as $\exp(i\theta)$. In this book, when the phase part θ is positive, we define it as the direction of the light propagation along the optical axis. Conversely, when the phase part θ is negative, the light propagates in the opposite direction. The property that the propagation direction can be determined by the sign of this phase part will be used later.

1.2 HOLOGRAPHY

Holography [14] was invented in 1947 by **Gabor** from Hungary. Holography was named using the Greek word "holos" meaning "all." This technique can record light waves emitted from three-dimensional objects in a two-dimensional medium (**hologram**), such as photographic dry plates, and can reconstruct the wavefront of the light from the recording medium. Both the amplitude and phase of the light wave are recorded on the hologram. Here, we briefly explain why holography can record and reconstruct three-dimensional images.

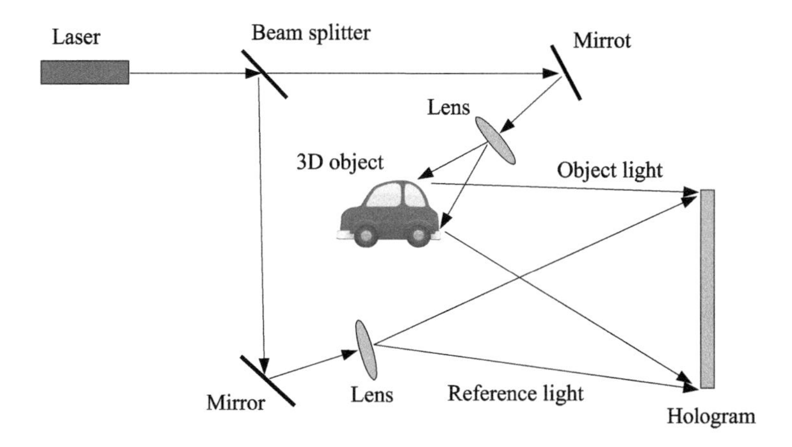

FIGURE 1.6 Hologram recording.

Figure 1.6 shows a typical optical system for hologram recording. Laser light is split into two beams by a beam splitter, one laser light is spread by a lens, and the laser is irradiated to a three-dimensional object to be recorded. The laser light diffused on the surface of the three-dimensional object reaches the hologram. This light is called **object light**. However, the three-dimensional information of the object cannot be recorded only with the object light. In holography, the other laser beam is spread by a lens and is irradiated to the hologram. This light is called **reference light**.

The hologram recording can be simplified as shown in Figure 1.7. The

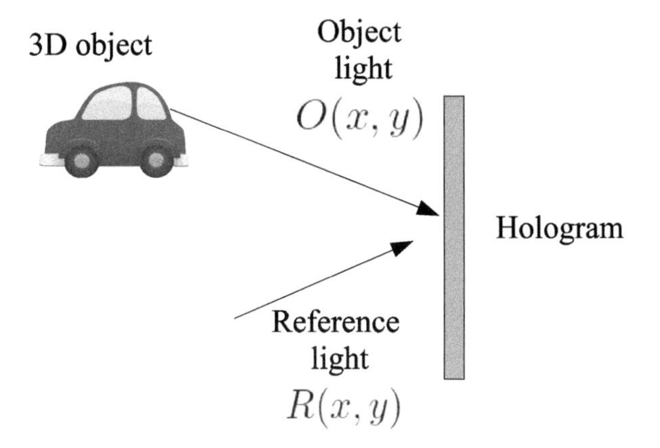

FIGURE 1.7 Simplified hologram recording.

object light $O(x, y)$ on the hologram plane is expressed as

$$O(x, y) = A(x, y) \exp(i\phi(x, y)), \tag{1.20}$$

where $A(x, y)$ and $\phi(x, y)$ are the amplitude and phase of the object. The reference light $R(x, y)$ is expressed as

$$R(x, y) = A_R(x, y) \exp(i\phi_R(x, y)), \tag{1.21}$$

where $A_R(x, y)$ and $\phi_R(x, y)$ are the amplitude and phase of the reference light.

Since the hologram is recorded as interference fringes of the object light and reference light, the hologram is written as

$$\begin{aligned} I(x, y) &= |O(x, y) + R(x, y)|^2 \\ &= \underbrace{|O(x, y)|^2 + |R(x, y)|^2}_{\text{Direct light}} + \underbrace{O(x, y)R^*(x, y)}_{\text{Object light}} + \underbrace{O^*(x, y)R(x, y)}_{\text{Conjugate light}}, \end{aligned} \tag{1.22}$$

where $*$ represents a **complex conjugate**.

Figure 1.8 is an optical system for reconstructing the three-dimensional image from the hologram. Three-dimensional images can be reconstructed by using the same reference light as when recording the hologram. This light is called **reconstruction light**.

The mathematical notation that illuminates the hologram $I(x, y)$ with the reconstruction light $R(x, y)$ is $I(x, y) \times R(x, y)$. Therefore, the reconstruction is expressed as

$$I(x,y)\times R(x,y) = \underbrace{R(x,y)(|O(x,y)|^2+|R(x,y)|^2)+}_{\text{Direct light}} \underbrace{O(x,y)}_{\text{Object light}} + \underbrace{O^*(x,y)R^2(x,y)}_{\text{Conjugate light}}. \tag{1.23}$$

The reconstruction light is diffracted by minute interference fringes of the hologram. Since a part of the diffracted light is exactly the same wavefront as the object light, an observer can recognize the recorded three-dimensional object by looking into the hologram.

In fact, the hologram simultaneously reconstructs **direct light**, **object light**, and **conjugate light** as shown in Figure 1.9. The object light is the desired light, the direct light passes directly through the hologram, and the conjugate light is a light conjugate to the object light. Although it is ideal that only the object light is reconstructed, the direct light and conjugate light are recorded as a penalty for recording three-dimensional information in a two-dimensional hologram.

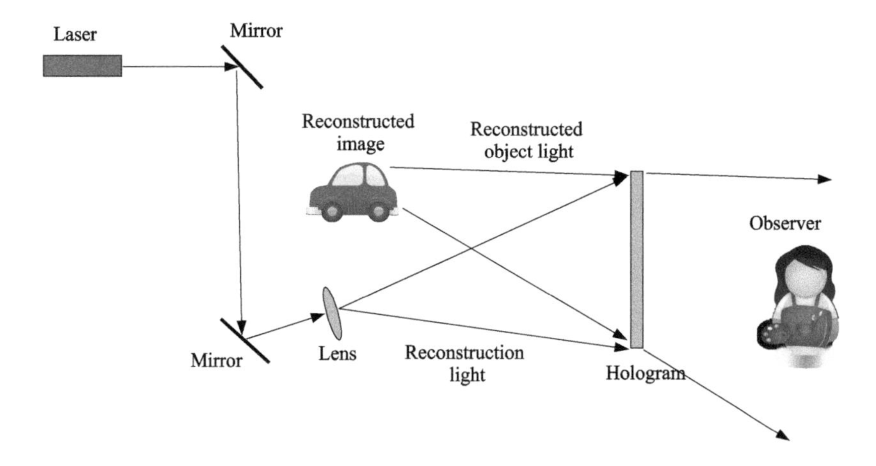

FIGURE 1.8 Reconstruction from a hologram.

1.2.1 INLINE HOLOGRAM AND OFF-AXIS HOLOGRAM

Depending on the incident angle of the reference light, the hologram is classified into two types: an **inline hologram** (also called an **on-axis hologram**) and an **off-axis hologram**. As shown on the left of Figure 1.10, an inline hologram is recorded when the incident angle of the reference light to the

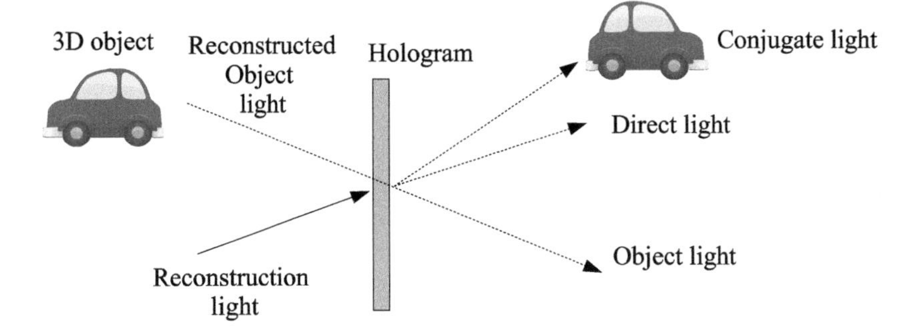

FIGURE 1.9 Reconstruction of direct, object and conjugate lights.

normal of the hologram is 0°. As shown on the right of Figure 1.10, an off-axis hologram is recorded with the reference light tilted by θ.

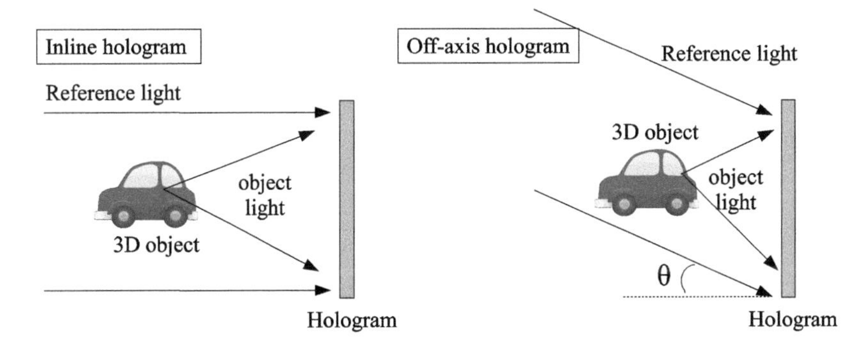

FIGURE 1.10 Inline hologram and off-axis hologram.

Let us express an inline hologram by mathematical expressions. Here, in the simplest case using a planar reference light with the amplitude of 1 and initial phase of 0, the reference light on the hologram plane is written as

$$R(x, y) = A_R(x, y) \exp(i\phi_R(x, y)) = 1. \tag{1.24}$$

Therefore, the inline hologram is expressed as

$$\begin{aligned} I(x, y) &= |O(x, y) + R(x, y)|^2 = |O(x, y) + 1|^2 \\ &= |O(x, y)|^2 + 1 + O(x, y) + O^*(x, y). \end{aligned} \tag{1.25}$$

The reconstruction from the inline hologram can be obtained by irradiating the reconstruction light, which is the same light as the reference light. The irradiation is mathematically expressed by multiplying the reconstruction light $R(x, y) = 1$ with the hologram I. The reconstruction can be written as

$$I(x, y) \times R(x, y) = |O(x, y) + R(x, y)|^2 = |O(x, y)|^2 + 1 + O(x, y) + O^*(x, y). \tag{1.26}$$

The first and second terms $|O(x,y)|^2 + 1$ is the direct light, the third term is the object light, and the forth term is the conjugate light. As mentioned in Section 1.1.2, the conjugate light represents the convergent light because the sign of the phase of the object light is inverted.

The object light, conjugate light, and direct light are reconstructed as shown in Figure 1.11. When the observer observes the hologram from the right side, the observer can see object light three-dimensionally in the back of the hologram. However, since the other unwanted lights are reconstructed simultaneously, it is very difficult to observe the wanted object light only.

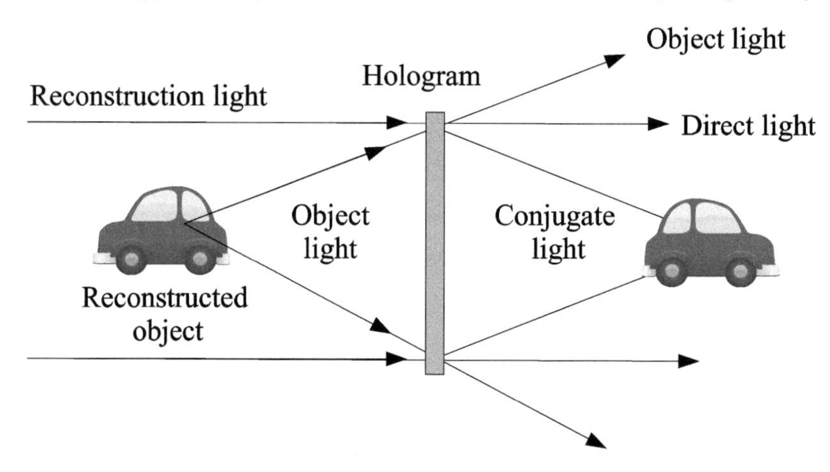

FIGURE 1.11 Reconstruction from an inline hologram.

Off-axis holograms can solve this problem. In a simplest case of the off-axis hologram, we use the reference light of a plane wave (amplitude 1, initial phase 0) with the incident angle θ, and we can write the reference light on the hologram plane as

$$R(x,y) = A_R(x,y)\exp(ikx\sin\theta) = \exp(ikx\sin\theta). \tag{1.27}$$

Therefore, the off-axis hologram can be expressed as

$$\begin{aligned} I(x,y) &= |O(x,y) + R(x,y)|^2 \\ &= |O(x,y)|^2 + 1 + O(x,y)\exp(-ikx\sin\theta) \\ &\quad + O^*(x,y)\exp(ikx\sin\theta). \end{aligned} \tag{1.28}$$

The reconstruction from the off-axis hologram is expressed as

$$\begin{aligned} I(x,y) \times R(x,y) &= (|O(x,y)|^2 + 1)\exp(ikx\sin\theta) + O(x,y) \\ &\quad + O^*(x,y)\exp(i2kx\sin\theta). \end{aligned} \tag{1.29}$$

The first term is the direct light and it propagates in the θ direction as shown in Figure 1.12. The second term is the object light. The third term is the conjugate light and it propagates in approximately the 2θ direction because of $\exp(i2kx \sin\theta) \approx \exp(ikx \sin 2\theta)$. [ii] Unlike an inline hologram, the object light, conjugate light and direct light are separated from each other. When an observer views the hologram from the right side, the conjugate light and direct light do not enter the observer's eyes and only the object light can be observed.

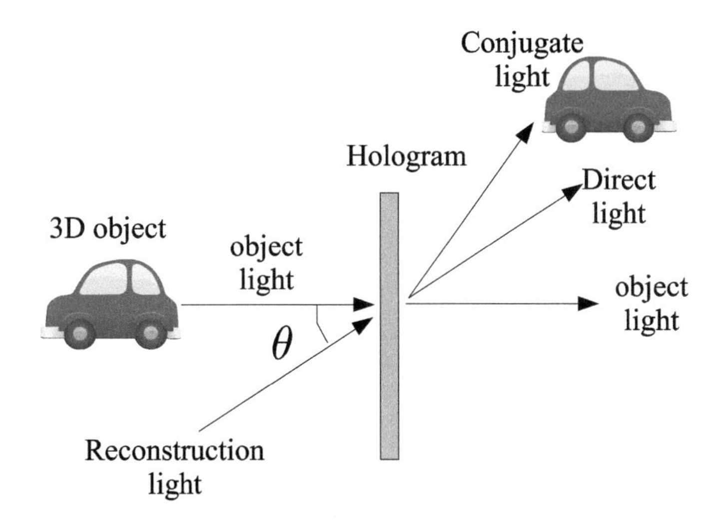

FIGURE 1.12 Reconstruction from an off-axis hologram.

1.2.2 TYPES OF HOLOGRAM

Holograms are classified according to various criteria. In the previous section, we classified inline holograms and off-axis holograms by the incident angle of the reference light. In this section, we briefly introduce types of holograms often used in computer holography. The classification in terms of the distance between a three-dimensional object and a hologram is as follows.

- Near distance : Image hologram
- Middle distance : Fresnel hologram
- Long distance: Fraunhofer hologram

A **Fresnel hologram** is a hologram for which the object light can be described by Fresnel diffraction. When the distance between the hologram and the object is very far, object light can be regarded as plane waves and be described by Fraunhofer diffraction. The hologram in this case is called a **Fraunhofer hologram**. The details of the Fresnel diffraction and Fraunhofer diffraction are explained in Chapter 2.

[ii] We used the approximation of $\sin 2\theta \approx 2\theta$.

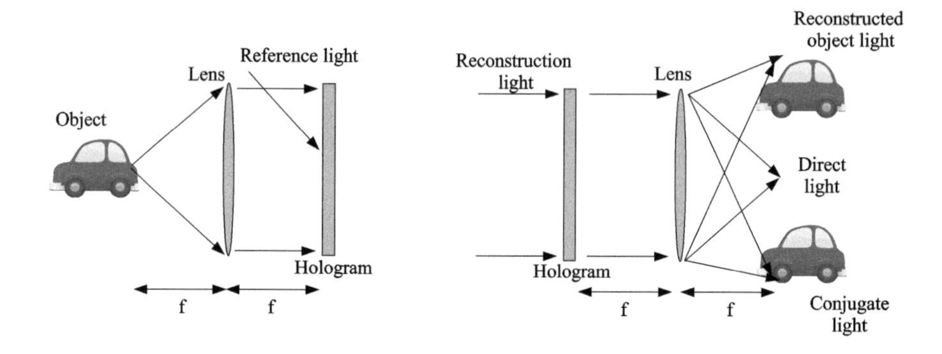

FIGURE 1.13 Fraunhofer hologram.

In practice, if it is difficult to place the object at far distance, Fraunhofer diffraction can be approximately realized by using a lens. If we place an object at the front focal position of the lens with the focal length f and a hologram at the back focal position as shown in the left of Figure 1.13, the object light can be regarded as Fraunhofer diffraction. Since the lens effect is calculated by Fourier transform, the hologram in Figure 1.13 is also called a **Fourier hologram**.

A Fourier hologram that does not use a lens is called a **lensless Fourier hologram**. As shown in Figure 1.14, the reference light is treated as a spherical wave through a pinhole. By placing the pinhole and the object at the same distance from the hologram, it is possible to record the Fourier hologram without using the lens.

On the other hand, a hologram recorded with an object placed near the hologram is called an **image hologram**. When an object has a certain size, it is difficult to actually place the object near the hologram. For example, as shown in Figure 1.15, an image hologram is recorded by the interference of the reference light and the object light formed near the hologram using a lens (focal length f).

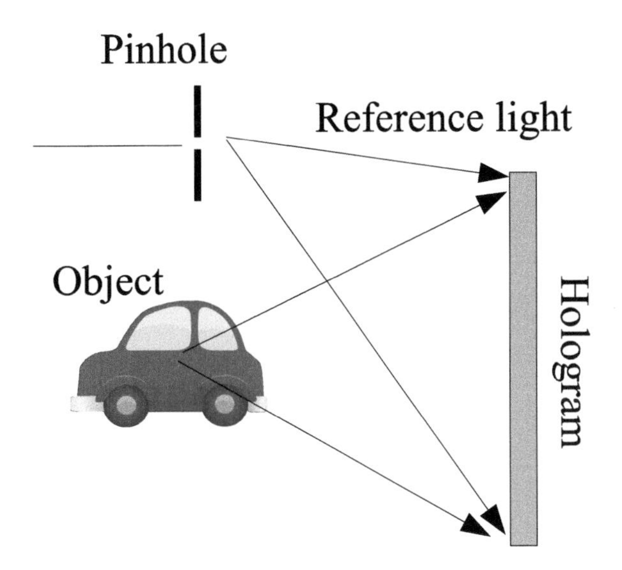

FIGURE 1.14 Lensless Fourier hologram.

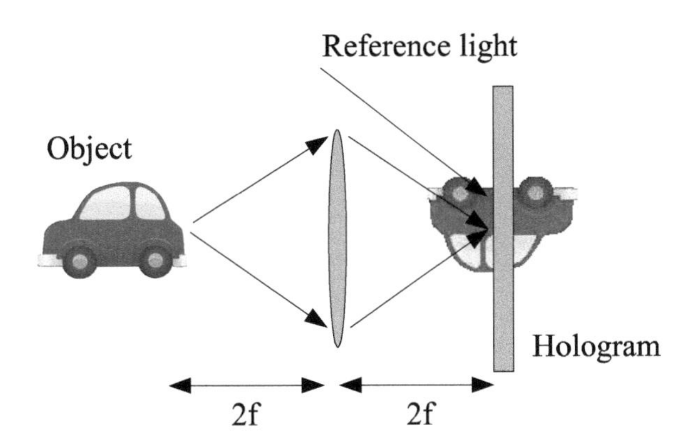

FIGURE 1.15 Image hologram.

2 Numerical diffraction calculation

Diffraction calculation is the most important calculation in computer holography. It is used for calculating light waves emitted from three-dimensional objects and used for reconstructing three-dimensional information recorded in holograms. Diffraction is mathematically expressed by diffraction integral. Diffraction calculation is derived from Maxwell's equation; however, this derivation is given in other good literature, for example, Ref. [13]. Starting from the Sommerfeld diffraction integral, we derive several diffraction calculations and explain their numerical forms. In addition, we illustrate how to implement numerical diffraction calculation using program language.

2.1 SOMMERFELD DIFFRACTION INTEGRAL

Diffraction integrals describe the propagation of light from a plane (x_1, y_1) to a plane (x_2, y_2), as shown in Figure 2.1. A light incident from the left side of Figure 2.1 is diffracted by $u_1(x_1, y_1)$, and then we observe the diffraction pattern on the plane (x_2, y_2). There are various diffraction integrals; however, using the **Sommerfeld diffraction integral** as a starting point, we derive various kinds of diffraction calculations often used in computer holography.

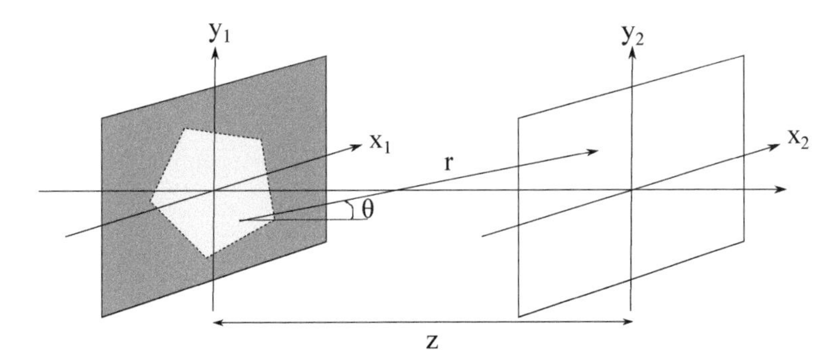

FIGURE 2.1 Diffraction integral.

The Sommerfeld diffraction integral is expressed as

$$u_2(x_2, y_2) = \frac{1}{i\lambda} \int\int u_1(x_1, y_1) \frac{\exp(ikr)}{r} \cos(\theta) dx_1 dy_1, \tag{2.1}$$

where $r = \sqrt{(x_2 - x_1)^2 + (y_2 - y_1)^2 + z^2}$ is the distance between (x_1, y_1) and (x_2, y_2), and $k = 2\pi/\lambda$ (λ is the wavelength) represents the wave number.

This expression can be considered as the superposition of spherical waves $\frac{\exp(ikr)}{r}$ emanating from each point of $u_1(x_1, y_1)$ to a point in $u_2(x_2, y_2)$. $1/\lambda$ attenuates the amplitude of the light, and $1/i = -i = \exp(-i\pi/2)$ expresses a phase delay of $\pi/2$. $\cos(\theta)$ is called an **inclination factor**, and as the angle θ formed by between r and the normal of u_1 in Figure 2.1 increases, the amplitude of the spherical wave $\frac{\exp(ikr)}{r}$ is attenuated.

Since the inclination factor can be written as $\cos(\theta) = z/r$, the Sommerfeld diffraction integral is also written as

$$u_2(x_2, y_2) = \frac{1}{i\lambda} \int\int u_1(x_1, y_1) \frac{\exp(ikr)}{r} \frac{z}{r} dx_1 dy_1. \tag{2.2}$$

2.2 ANGULAR SPECTRUM METHOD (PLANAR WAVE DECOMPOSITION)

The **angular spectrum method** (also called **planar wave decomposition**) is a diffraction calculation often used in computer holography and can be derived from Eq. (2.2). To derive the angular spectrum method, we use the **convolution theorem**[i] which is one of the Fourier transform theorems.

In this book, the two-dimensional **Fourier transform**[ii] is defined as

$$U(f_x, f_y) = \int_{-\infty}^{\infty} \int_{-\infty}^{\infty} u(x, y) \exp(-i2\pi(f_x x + f_y y)) dx dy = \mathcal{F}\left[u(x, y)\right], \tag{2.3}$$

and the two-dimensional inverse Fourier transform is defined as

$$u(x, y) = \int_{-\infty}^{\infty} \int_{-\infty}^{\infty} U(f_x, f_y) \exp(i2\pi(f_x x + f_y y)) df_x df_y = \mathcal{F}^{-1}\left[U(f_x, f_y)\right]. \tag{2.4}$$

$\mathcal{F}[\cdot]$ and $\mathcal{F}^{-1}[\cdot]$ are operators representing the Fourier transform and inverse Fourier transform, and (f_x, f_y) denotes coordinates in the frequency domain.

Generally, the convolution integral has the following form (Eq. (2.5)) and can be calculated by using the Fourier transforms according to the **convolution theorem.**[iii]

$$\begin{aligned} u_2(x_2, y_2) &= \int_{-\infty}^{\infty} \int_{-\infty}^{\infty} u_1(x_1, y_1) h(x_2 - x_1, y_2 - y_1) dx_1 dy_1 && (2.5) \\ &= u_1(x_1, y_1) \otimes h(x_1, y_1) && (2.6) \\ &= \mathcal{F}^{-1}\left[\mathcal{F}\left[u_1(x_1, y_1)\right] \mathcal{F}\left[h(x_1, y_1)\right]\right] && (2.7) \\ &= \mathcal{F}^{-1}\left[\mathcal{F}\left[u_1(x_1, y_1)\right] H(f_x, f_y)\right] && (2.8) \end{aligned}$$

[i] See Chapter 7.

[ii] Some definitions require $1/2\pi$ or $1/\sqrt{2\pi}$ before the integration of the Fourier transform and inverse Fourier transform, but ignore them in this book.

[iii] See Chapter 7.

$h(x_1, y_1)$ is called the **impulse response**, which is the response of a linear system when a delta function is input to the linear system, and $\otimes$ is the operator of the convolution integral. $H(f_x, f_y) = \mathcal{F}\left[h(x_1, y_1)\right]$ is called the **transfer function**.

When the convolution theorem is applied to Eq. (2.2), we obtain

$$u_2(x_2, y_2) = \mathcal{F}^{-1}\left[\mathcal{F}\left[u_1(x_1, y_1)\right]\mathcal{F}\left[\frac{z}{i\lambda}\frac{\exp(ikr)}{r^2}\right]\right]. \tag{2.9}$$

The transfer function $H(f_x, f_y) = \mathcal{F}\left[\frac{z}{i\lambda}\frac{\exp(ikr)}{r^2}\right]$ can be obtained analytically [15,16],

$$H(f_x, f_y) = \mathcal{F}\left[\frac{z}{i\lambda}\frac{\exp(ikr)}{r^2}\right] = \exp\left(i2\pi z\sqrt{\frac{1}{\lambda^2} - f_x^2 - f_y^2}\right). \tag{2.10}$$

Eventually, the angular spectrum method is expressed by

$$\begin{aligned} u_2(x_2, y_2) &= \mathcal{F}^{-1}\left[\mathcal{F}\left[u_1(x_1, y_1)\right]\exp\left(i2\pi z\sqrt{\frac{1}{\lambda^2} - f_x^2 - f_y^2}\right)\right] \\ &= \mathcal{F}^{-1}\left[U(f_x, f_y)\exp\left(i2\pi z\sqrt{\frac{1}{\lambda^2} - f_x^2 - f_y^2}\right)\right], \end{aligned} \tag{2.11}$$

where

$$\begin{aligned} U(f_x, f_y) &= \mathcal{F}\left[u_1(x_1, y_1)\right] \\ &= \int_{-\infty}^{\infty}\int_{-\infty}^{\infty} u_1(x_1, y_1)\exp(-2\pi i(f_x x_1 + f_y y_1))dx_1 dy_1. \end{aligned} \tag{2.12}$$

is called the **angular spectrum**. The physical meaning of the angular spectrum $U(f_x, f_y)$, which is described in detail in Section 2.2.1, is a plane wave propagating at an angle corresponding to f_x, f_y.

If Eq. (2.11) is $f_x^2 + f_y^2 > \frac{1}{\lambda^2}$ then $\exp\left(-2\pi z\sqrt{f_x^2 + f_y^2 - \frac{1}{\lambda^2}}\right)$. This is called **evanescent light** (near-field light), and is attenuated exponentially as z increases because it does not include imaginary unit i. This light is used for, for example, the super-resolution of fluorescent microscopes, but it is not dealt with in this book.

The angular spectrum method is important because it can be calculated by using the Fourier transforms without applying any approximation to Eq. (2.2). When computing the angular spectrum method with a computer, it is possible to use the fast Fourier transform, which is advantageous in terms of calculation time.

2.2.1 INTERPRETATION OF ANGULAR SPECTRUM

The angular spectrum method is also important because it gives a physical interpretation that diffraction calculation can be expressed by the superposition of plane waves. When taking the inverse Fourier transform of the angular spectrum (Eq. (2.12)), it can be written as

$$\begin{aligned} u_1(x_1, y_1) &= \mathcal{F}^{-1}\Big[U(f_x, f_y)\Big] \\ &= \int_{-\infty}^{\infty}\int_{-\infty}^{\infty} U(f_x, f_y)\exp(2\pi i(f_x x_1 + f_y y_1))df_x df_y. \end{aligned} \quad (2.13)$$

From Eq. (1.11), the plane wave with the amplitude a traveling to the wave vector direction $\mathbf{k} = (k_x, k_y, k_z) = k(\alpha, \beta, \gamma)$ (k: the wave number) can be written as

$$\begin{aligned} u(x, y, z) &= a\exp(i\mathbf{k}\cdot\mathbf{x}), \\ &= a\exp(i(k_x x + k_y y + k_z z)). \end{aligned} \quad (2.14)$$

We define $u(x_1, y_1, 0) = u(x_1, y_1)$ at the position of $(x_1, y_1, 0)$. Comparing Eq. (2.13) with Eq. (2.14), $u_1(x_1, y_1)$ is understood to be expressed as the sum of plane waves with various spatial frequencies whose amplitude is the angular spectrum $U(f_x, f_y)$. We have the following relation:

$$u(x_1, y_1) = U(f_x, f_y)\exp(2\pi i(f_x x_1 + f_y y_1)) = a\exp(i(k_x x_1 + k_y y_1)). \quad (2.15)$$

Therefore, the wave vector and the spatial frequency have the following relation

$$\begin{aligned} \alpha &= \lambda f_x, \\ \beta &= \lambda f_y, \\ \gamma &= \sqrt{1 - (\lambda f_x)^2 - (\lambda f_y)^2}, \end{aligned} \quad (2.16)$$

where the third equation is derived from the relation $\mathbf{k} = |k| = k\sqrt{k_x^2 + k_y^2 + k_z^2}$.

Eq. (2.11) is written as

$$\begin{aligned} u_2(x_2, y_2) &= \mathcal{F}^{-1}\left[U(f_x, f_y)\exp\left(i2\pi z\sqrt{\frac{1}{\lambda^2} - f_x^2 - f_y^2}\right)\right] \\ &= \int\int U(f_x, f_y)\exp\left(i2\pi z\sqrt{\frac{1}{\lambda^2} - f_x^2 - f_y^2}\right) \\ &\quad \exp(2\pi i(f_x x_2 + f_y y_2))df_x df_y \\ &= \int\int U(f_x, f_y)\exp(ikz\gamma) \\ &\quad \times\exp(2\pi i(f_x x_2 + f_y y_2))df_x df_y. \end{aligned} \quad (2.17)$$

This suggests that the sum of the plane waves represented by the angular spectrum of $u_1(x_1, y_1)$ reaches the (x_2, y_2) plane is $u_2(x_2, y_2)$. $\exp(ikz\gamma)$ has the effect of converting the angular spectrum U of the (x_1, y_1) plane into an angular spectrum at the position away from z. This is why the angular spectrum method is called planar wave decomposition.

2.3 FRESNEL DIFFRACTION

Fresnel diffraction plays an important role in computer holography as well as the angular spectrum method. Fresnel diffraction can be derived by approximating Eq. (2.2). Since the angular spectrum method does not have an approximation, it seems that it can be used in any case; however, in the actual numerical calculation, the angular spectrum method is used in a short propagation distance, and Fresnel diffraction is in a long propagation distance. See Section 2.6.4 for details.

The distance r in the exponential term of Eq. (2.2) is approximated by **Taylor expansion** ($\sqrt{1+a} \approx 1 + \frac{1}{2}a - \frac{1}{8}a^2 + \cdots$. This expression is also called **binomial expansion**) and is written as

$$\begin{aligned} r &= z\sqrt{1 + \frac{(x_2 - x_1)^2 + (y_2 - y_1)^2}{z^2}}, \\ &\approx z + \frac{(x_2 - x_1)^2 + (y_2 - y_1)^2}{2z} - \frac{((x_2 - x_1)^2 + (y_2 - y_1)^2)^2}{8z^3} + \cdots, \end{aligned} \tag{2.18}$$

In the field of optics, **Fresnel approximation** (or **paraxial approximation**) approximates the distance r using up to second-order terms of this approximation. Since the distance r in the exponential term of Eq. (2.2) is related to the phase of the light wave, it needs to be computed relatively accurately by Eq. (2.18). However, the other r in Eq. (2.2) affects only the amplitude of the light wave, so that roughly $r \approx z$ is acceptable.

By applying these approximations to Eq. (2.2), we obtain

$$\begin{aligned} u_2(x_2, y_2) &= \frac{1}{i\lambda} \int\int u_1(x_1, y_1) \frac{\exp(ikr)}{r} \frac{z}{r} dx_1 dy_1 \\ &\approx \frac{1}{i\lambda} \int\int u_1(x_1, y_1) \frac{\exp(ik(z + \frac{(x_2-x_1)^2+(y_2-y_1)^2}{2z}))}{z} \frac{z}{z} dx_1 dy_1. \end{aligned} \tag{2.19}$$

Eventually, Fresnel diffraction can be written as

$$\begin{aligned} u_2(x_2, y_2) &= \frac{\exp(i\frac{2\pi}{\lambda}z)}{i\lambda z} \int\int_{-\infty}^{+\infty} u_1(x_1, y_1) \\ &\quad \times \exp(i\frac{\pi}{\lambda z}((x_2 - x_1)^2 + (y_2 - y_1)^2)) dx_1 dy_1. \end{aligned} \tag{2.20}$$

In order for this approximation to hold, the third term of Eq. (2.18) must be sufficiently smaller than λ, that is,

$$\lambda \gg \frac{((x_2 - x_1)^2 + (y_2 - y_1)^2)^2}{8z^3}. \tag{2.21}$$

For example, if the wavelength is $\lambda = 600$ nm, and the maximum range is $|x_2 - x_1|_{max} = 1$ cm and $|y_2 - y_1|_{max} = 1$ cm, Fresnel approximation will be well held in $z \gg 0.2$ m because of

$$\begin{aligned} z^3 \gg \frac{((1[\text{cm}])^2 + (1[\text{cm}])^2)^2}{8 \times 600[\text{nm}]} &= \frac{4 \times 10^{-8}}{4.8 \times 10^{-6}}[\text{m}^3] \\ &= \frac{10^{-2}}{1.2}[\text{m}^3]. \end{aligned} \tag{2.22}$$

When performing Fresnel diffraction with a computer, in many cases, Eq. (2.20) is not directly numerically integrated in terms of the calculation time. It is often calculated in one of the ways to be introduced in the next sections.

2.3.1 FRESNEL DIFFRACTION: CONVOLUTION FORM

This type of Fresnel diffraction is expressed by the following equation using the convolution theorem as well as the angular spectrum method:

$$\begin{aligned} u_2(x_2, y_2) &= \frac{\exp(i\frac{2\pi}{\lambda}z)}{i\lambda z} \iint_{-\infty}^{+\infty} u_1(x_1, y_1) \\ &\quad \times \exp(i\frac{\pi}{\lambda z}((x_2 - x_1)^2 + (y_2 - y_1)^2))dx_1 dy_1, \\ &= \frac{\exp(i\frac{2\pi}{\lambda}z)}{i\lambda z} \times u_1(x_1, y_1) \otimes \exp(i\frac{\pi}{\lambda z}(x_1^2 + y_1^2)), \qquad (2.23) \\ &= \frac{\exp(i\frac{2\pi}{\lambda}z)}{i\lambda z} \mathcal{F}^{-1}\Big[\mathcal{F}\Big[u_1(x_1, y_1)\Big] \cdot \mathcal{F}\Big[h_f(x_1, y_1)\Big]\Big], \qquad (2.24) \end{aligned}$$

where $h_f(x, y)$ is the **impulse response** defined as

$$h_f(x, y) = \exp(i\frac{\pi}{\lambda z}(x^2 + y^2)). \tag{2.25}$$

This equation can be calculated by three Fourier transforms.

The Fourier transform of the multiplication of this impulse response $h_f(x, y)$ with $\frac{\exp(i\frac{2\pi}{\lambda}z)}{i\lambda z}$ is analytically solved as

$$\begin{aligned} H_f(f_x, f_y) &= \mathcal{F}\Big[\frac{\exp(i\frac{2\pi}{\lambda}z)}{i\lambda z} h_f(x, y)\Big], \\ &= \exp(i\frac{2\pi}{\lambda}z)\exp(-i\pi\lambda z(f_x^2 + f_y^2)), \end{aligned} \tag{2.26}$$

where f_x, f_y are the coordinates in the frequency domain. In this case, the convolution type Fresnel diffraction can be written as

$$u_2(x_2, y_2) = \mathcal{F}^{-1}\left[\mathcal{F}\left[u_1(x_1, y_1)\right] H_f(f_x, f_y)\right]. \tag{2.27}$$

This equation can be calculated by two Fourier transforms.

2.3.2 FRESNEL DIFFRACTION: FOURIER TRANSFORM FORM

Fresnel diffraction in the Fourier transform form is expressed by the following equation using the Fourier transform once:

$$\begin{aligned} u_2(x_2, y_2) &= \frac{\exp(i\frac{2\pi}{\lambda}z)}{i\lambda z} \iint_{-\infty}^{+\infty} u_1(x_1, y_1) \\ &\quad \times \exp(i\frac{\pi}{\lambda z}((x_2 - x_1)^2 + (y_2 - y_1)^2))dx_1 dy_1, \\ &= \frac{\exp(i\frac{2\pi}{\lambda}z)}{i\lambda z} \iint_{-\infty}^{+\infty} u_1(x_1, y_1) \\ &\quad \times \exp(i\frac{\pi}{\lambda z}(x_2^2 - 2x_2x_1 + x_1^2 + y_2^2 - 2y_2y_1 + y_1^2))dx_1 dy_1, \\ &= \frac{\exp(i\frac{2\pi}{\lambda}z)}{i\lambda z} \exp(i\frac{\pi}{\lambda z}(x_2^2 + y_2^2)) \\ &\quad \times \iint_{-\infty}^{+\infty} u_1(x_1, y_1) \exp(i\frac{\pi}{\lambda z}(x_1^2 + y_1^2)) \\ &\quad \times \exp(-2\pi i(\frac{x_2 x_1}{\lambda z} + \frac{y_2 y_1}{\lambda z}))dx_1 dy_1. \end{aligned} \tag{2.28}$$

Here we define $u_1'(x_1, y_1)$ as

$$u_1'(x_1, y_1) = u_1(x_1, y_1) \exp(i\frac{\pi}{\lambda z}(x_1^2 + y_1^2)), \tag{2.29}$$

and replace $\frac{x_1}{\lambda z}$ and $\frac{y_1}{\lambda z}$ with the new variables x_1' and y_1'. Therefore, Eq. (2.28) can be rewritten as

$$\begin{aligned} u_2(x_2, y_2) &= \frac{\exp(i\frac{2\pi}{\lambda}z)}{i\lambda z} \exp(i\frac{\pi}{\lambda z}(x_2^2 + y_2^2)) \\ &\quad \underbrace{\iint_{-\infty}^{+\infty} u_1'(x_1, y_1) \exp(-2\pi i(x_2 x_1' + y_2 y_1'))dx_1 dy,}_{\text{This is the same as Fourier transform}} \\ &= \frac{\exp(i\frac{2\pi}{\lambda}z)}{i\lambda z} \exp(i\frac{\pi}{\lambda z}(x_2^2 + y_2^2))\mathcal{F}\left[u_1'(x_1', y_1')\right]. \end{aligned} \tag{2.30}$$

This Fresnel diffraction can be calculated with only a single Fourier transform.

2.4 FRAUNHOFER DIFFRACTION

Fraunhofer diffraction is used for the calculation of the far diffraction region. In this diffraction calculation, if the phase of the exponential term in Eq. (2.29) in the Fresnel diffraction of the Fourier transform form is $\frac{\pi}{\lambda z}(x_1^2 + y_1^2) \ll 2\pi$, it can be approximated[iv] as

$$\exp(i\frac{\pi}{\lambda z}(x_1^2 + y_1^2)) \approx 1. \tag{2.31}$$

From Eq. (2.30), Fraunhofer diffraction can be written as

$$u_2(x_2, y_2) = \frac{\exp(i\frac{2\pi}{\lambda}z)}{i\lambda z}\exp(i\frac{\pi}{\lambda z}(x_2^2 + y_2^2))\mathcal{F}\left[u_1(x_1, y_1)\right]. \tag{2.32}$$

z where Fraunhofer diffraction holds is

$$z \gg \frac{x_1^2 + y_1^2}{2\lambda}. \tag{2.33}$$

For example, if the wavelength is $\lambda = 600$ nm, and the maximum range is $|x_1|_{max} = 1$ cm and $|y_1|_{max} = 1$ cm, Fraunhofer diffraction is well held under

$$z \gg \frac{(1[\text{cm}])^2 + (1[\text{cm}])^2}{2 \times 600[\text{nm}]} = \frac{2 \times 10^{-4}}{1.2 \times 10^{-6}}[\text{m}] \approx 167[\text{m}]. \tag{2.34}$$

However, because $z \gg 167$ m is too far, it is difficult to observe the Fraunhofer diffraction in an actual optical system. Since it is known that the diffraction pattern of the focal plane of a lens is Fraunhofer diffraction, lenses are often used in experiments. If we set the range to $|x_1| = 0.5$ mm and $|y_1| = 0.5$ mm, we get $z \gg 0.4$ m so that we can observe Fraunhofer diffraction without using a lens.

2.5 OPERATOR OF DIFFRACTION

In holography, diffraction calculation is heavily used, so it is redundant to describe the integral equations of diffraction every time. Introducing an **operator** representing the diffraction calculation can simplify the notation. Although there are no unified notations of operators, we define $\text{Prop}_z[\cdot]$ as the operator in this book. For example, Fresnel diffraction propagating at the distance z is expressed as

$$\begin{aligned} u_2(x_2, y_2) &= \text{Prop}_z[u_1(x_1, y_1)], \\ &= \frac{\exp(i\frac{2\pi}{\lambda}z)}{i\lambda z}\iint_{-\infty}^{+\infty} u_1(x_1, y_1) \\ &\quad \times \exp(i\frac{\pi}{\lambda z}((x_2 - x_1)^2 + (y_2 - y_1)^2))dx_1 dy_1. \end{aligned} \tag{2.35}$$

[iv]The phase change is much smaller than 2π (almost 0 radian).

It is also useful to know that the diffraction operator has the following properties:

$$\mathrm{Prop}_z[u_a(x,y)+u_b(x,y)] = \mathrm{Prop}_z[u_a(x,y)]+\mathrm{Prop}_z[u_b(x,y)], \tag{2.36}$$

$$\mathrm{Prop}_z[c\,u(x,y)] = c\mathrm{Prop}_z[u(x,y)], \tag{2.37}$$

$$\mathrm{Prop}_{z_1+z_2}[u(x,y)] = \mathrm{Prop}_{z_2}[\mathrm{Prop}_{z_1}[u(x,y)]]. \tag{2.38}$$

The first and the second equations express the linearity of the operator.

2.6 NUMERICAL DIFFRACTION CALCULATION

This section explains numerical implementations written in C language for Fresnel diffraction in the integral form (Eq. (2.20)) introduced in Section 2.3 and the convolution form (Eq. (2.24)). Equation (2.20) and Eq. (2.24) are simple expressions, but the following points need to be kept in mind when implementing them.

- Discretization
- Handling complex numbers in programs
- Zero padding
- Fast Fourier transform and quadrant exchange

2.6.1 DISCRETIZATION

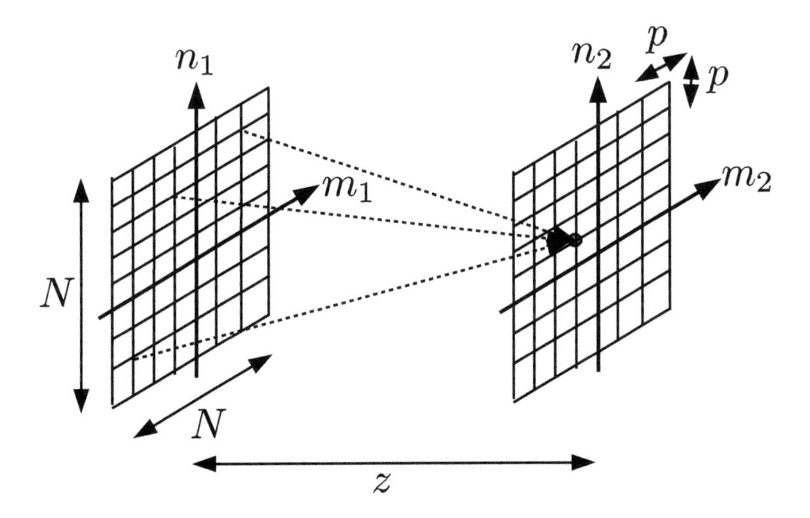

FIGURE 2.2 Discretization of Fresnel diffraction.

First, we consider the implementation of Fresnel diffraction of integral form (Eq. (2.20)). We write it again as

$$u_2(x_2, y_2) = \frac{\exp(i\frac{2\pi}{\lambda}z)}{i\lambda z}\int_{-\infty}^{+\infty}\int_{-\infty}^{+\infty} u_1(x_1, y_1) \times \exp(i\frac{\pi}{\lambda z}((x_2 - x_1)^2 + (y_2 - y_1)^2))dx_1 dy_1. \quad (2.39)$$

Since the coordinates (x_1, y_1) and (x_2, y_2) take continuous values, it is necessary to **discretize** as shown in Figure 2.2.

If both the source and the destination planes are sampled with $N \times N$ pixels, the coordinates can be written as

$$x_1 = pm_1, \quad y_1 = pn_1, \quad x_2 = pm_2, \quad y_2 = pn_2, \quad (2.40)$$

where p is the **sampling interval** (also referred to as the **sampling pitch**). m_1, n_1, m_2, n_2 are discretized coordinates and take the following integer values:

$$-\frac{N}{2} \le m_1 \le \frac{N}{2} - 1, \quad -\frac{N}{2} \le n_1 \le \frac{N}{2} - 1,$$
$$-\frac{N}{2} \le m_2 \le \frac{N}{2} - 1, \quad -\frac{N}{2} \le n_2 \le \frac{N}{2} - 1.$$

In this case, the same sampling interval is used in both the vertical and horizontal directions. The integral symbol $\int_{-\infty}^{+\infty}$ can be replaced by $\sum_{-N/2}^{N/2-1}$ and $dx_1 dy_1 \approx p^2$, so the discretized Fresnel diffraction is expressed as

$$u_2(m_2, n_2) = \frac{p^2 \exp(i\frac{2\pi}{\lambda}z)}{i\lambda z}\sum_{n_1=-\frac{N}{2}}^{\frac{N}{2}-1}\sum_{m_1=-\frac{N}{2}}^{\frac{N}{2}-1} u_1(m_1, n_1) \times \exp(\frac{i\pi p^2}{\lambda z}((m_2 - m_1)^2 + (n_2 - n_1)^2). \quad (2.41)$$

In order to calculate a point (m_2, n_2) in the destination plane in Figure 2.2, it is necessary to add all the light emitted from each sampling point in the source plane. If we want to observe the diffraction intensity pattern, we can just take the absolute square of $u_2(m_2, n_2)$. At this time, since the coefficient of Eq. (2.41) treats z as a constant at a certain distance and $|\frac{p^2 \exp(i\frac{2\pi}{\lambda}z)}{i\lambda z}|^2$ is a constant real value, we can omit the calculation of $|\frac{p^2 \exp(i\frac{2\pi}{\lambda}z)}{i\lambda z}|^2$.

Note that $u_1(m_1, n_1)$ and $u_2(m_2, n_2)$ are complex numbers. When handling complex numbers in a computer program, we handle real and imaginary parts of the complex numbers as separate variables. There are various definitions, but in this book we use the complex type (fftw_complex) defined in the fast Fourier transform library[v] described later. This complex type is defined by an

[v]FFTW

array of two elements, where the index 0 of the array is the real part and the index 1 is the imaginary part (Listing 2.1).

Listing 2.1 Definition of a complex number.

```
1 typedef double fftw_complex[2];
```

Equation (2.41) also requires complex multiplication of $u_1(m_1, n_1)$ and $\exp(\cdot)$. The complex multiplication of the complex number $c = a \times b$ is written as shown in Listing 2.2.

Listing 2.2 Complex multiplication.

```
1 fftw_complex a,b,c;
2 c[0] += a[0] * b[0] - a[1] * b[1];
3 c[1] += a[0] * b[1] + a[1] * b[0];
```

Therefore, the source code of the Fresnel diffraction which directly computes Figure 2.2 can be written as shown in Listing 2.3.

Listing 2.3 Fresnel diffraction.

```
1  fftw_complex* fresnel_direct(
2    fftw_complex *u, int N, double lambda, double z, double p)
3  {
4    fftw_complex *u2 =
5      (fftw_complex*)malloc(sizeof(fftw_complex)*N*N);
6
7    for (int n2 = 0; n2 < N; n2++){
8      for (int m2 = 0; m2 < N; m2++){
9        fftw_complex tmp;
10       tmp[0] = 0.0;
11       tmp[1] = 0.0;
12       for (int n1 = 0; n1 < N; n1++){
13         for (int m1 = 0; m1 < N; m1++){
14           int idx1 = m1 + n1*N;
15           double dx = ((m2 - N / 2) - (m1 - N / 2))*p;
16           double dy = ((n2 - N / 2) - (n1 - N / 2))*p;
17           double phase = (dx*dx + dy*dy)*M_PI / (lambda*z);
18           fftw_complex e, t;
19           e[0] = cos(phase);
20           e[1] = sin(phase);
21           t[0] = u[idx1][0];
22           t[1] = u[idx1][1];
23           tmp[0] += t[0] * e[0] - t[1] * e[1];
24           tmp[1] += t[0] * e[1] + t[1] * e[0];
25         }
26       }
27       int idx2 = m2 + n2*N;
28       u2[idx2][0] = tmp[0];
29       u2[idx2][1] = tmp[1];
```

```
30      }
31    }
32
33    free(u);
34    return u2;
35  }
```

The function "fresnel_direct" performs the Fresnel diffraction using the array "u" storing the source plane, and the diffracted result is stored in the array "u2." The arguments lambda, z, and p set the parameters of wavelength, propagation distance, and sampling interval, respectively.[vi] We subtract $N/2$ from the loop variables m_2, n_2, m_1, n_1 which represent the coordinates, because we place the origin at the center of the calculation area.

This implementation requires a quadruple loop. In numerical calculation, the number of loops greatly affects the calculation time. In the case of $N \times N$ pixels, it takes a very long time to calculate the entire propagation because it requires a loop count proportional to N^4.

When we want to observe the diffraction intensity pattern, we just take the absolute square of the resultant of $u_2(m_2, n_2)$ using Listing 2.4. The light intensity is stored in the real part of the complex array u.

Listing 2.4 Intensity.

```
1  void intensity(fftw_complex *u, int N)
2  {
3    for (int i = 0; i<N*N; i++){
4      double re = u[i][0];
5      double im = u[i][1];
6      u[i][0]=re*re + im*im;
7    }
8  }
```

2.6.2 IMPLEMENTATION OF FRESNEL DIFFRACTION USING FFT

Next, we consider the implementation of the Fresnel diffraction (Eq. (2.24)) in the convolution form using Fourier transform. We write the equation again as

$$\begin{aligned} u_2(x_2, y_2) &= \frac{\exp(i\frac{2\pi}{\lambda}z)}{i\lambda z} \times u_1(x_1, y_1) \otimes \exp(i\frac{\pi}{\lambda z}(x_1^2 + y_1^2)) \\ &= \frac{\exp(i\frac{2\pi}{\lambda}z)}{i\lambda z} \mathcal{F}^{-1}\left[\mathcal{F}\left[u_1(x_1, y_1)\right] \cdot \mathcal{F}\left[h(x_1, y_1)\right]\right], \end{aligned} \tag{2.42}$$

[vi]We use M_PI defined in C standard math library "math.h" as π.

where the impulse response $h(x, y)$ is defined as

$$h(x, y) = \exp(i\frac{\pi}{\lambda z}(x^2 + y^2)). \tag{2.43}$$

Although this equation is the convolution integral, it can be finally described by the Fourier transforms. The **fast Fourier transform (FFT)** is a well-known algorithm for efficiently calculating the Fourier transform. The computational complexity for one-dimensional FFT when the sample number is N points is $N \log_2 N$. In the case of two-dimensional FFT ($N \times N$ points), the computational complexity is $N^2 \log_2 N$.

In the Fresnel diffraction of Eq. (2.39) with $N \times N$ sampling points, computational complexity proportional to N^4 was necessary, whereas the Fresnel diffraction using FFTs can greatly reduce the computational complexity. The ratio of the computational complexity with/without FFTs is

$$\frac{N^2 \log_2 N}{N^4} = \frac{\log_2 N}{N^2}. \tag{2.44}$$

For example, when $N = 1,024 (= 2^{10})$, the ratio of the computational complexity is

$$10/1024^2 \approx 1/10^5. \tag{2.45}$$

Therefore, the calculation amount decreases to about 1 / 100,000.

It is necessary to implement the convolution calculation using FFTs by paying attention to the following points.

- Circular convolution and linear convolution
- Periodicity of FFT
- The position of the low-frequency component and the high-frequency component of the spectrum obtained by FFT

CIRCULAR CONVOLUTION AND LINEAR CONVOLUTION

For simplicity of discussion, we consider one-dimensional convolution for $u_1(m_1)$ and $h(m_1)$ using FFTs. The convolution is expressed as

$$u_2(m_2) = u_1(m_1) \otimes h(m_1) \tag{2.46}$$

$$= \sum_{m_1=-N/2}^{N/2} u_1(m_1) h(m_2 - m_1) \tag{2.47}$$

$$= \mathrm{FFT}^{-1}\left[\mathrm{FFT}\left[u_1(m_1)\right] \mathrm{FFT}\left[h(m_1)\right]\right]. \tag{2.48}$$

If we directly compute the convolution of Figure 2.3(a) and Figure 2.3(b) using Eq. (2.47), we obtain Figure 2.3(c). This is called **linear convolution**, and we can obtain the correct result.

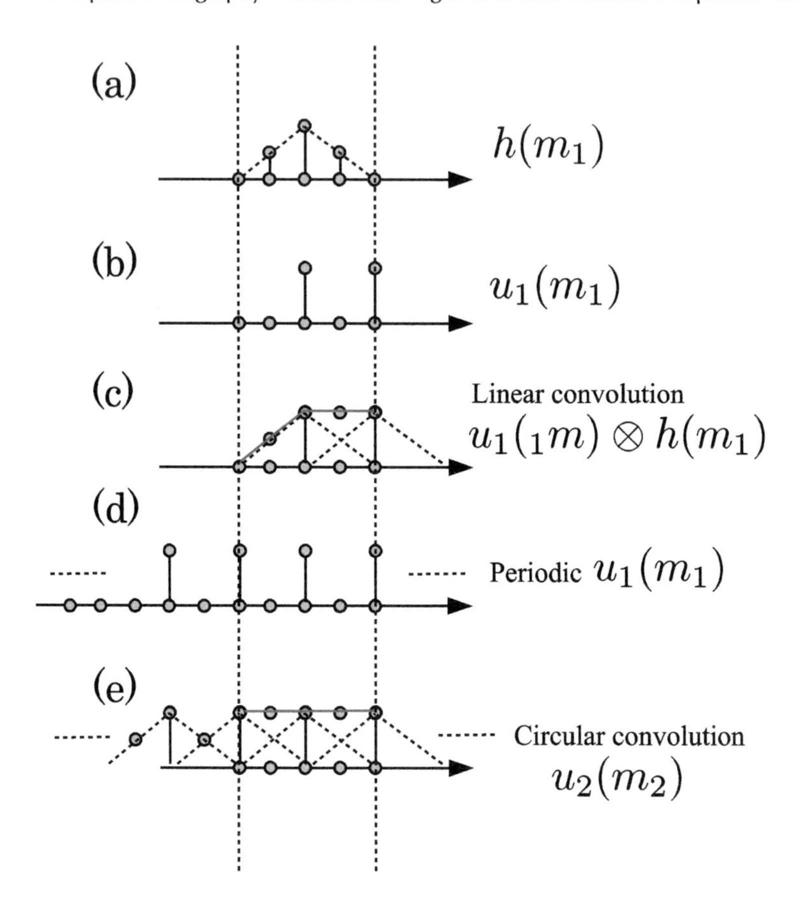

FIGURE 2.3 Linear convolution and circular convolution.

On the other hand, we need to be careful when the convolution is performed with FFTs (Eq. (2.48)). Although the FFT is an algorithm for the Fourier transform, in fact, it is similar to a Fourier series. Therefore, $u_1(m_1)$ and $h(m_1)$ are regarded as periodic signals (Figure 2.3(d) shows periodic $u_1(m_1)$). When convolving the periodic $u_1(m_1)$ and $h(m_1)$, the border of $h(m_1)$ appears on the opposite side of the calculation result as shown in Figure 2.3(e). Such a convolution is called **circular convolution**, and the wraparound is superimposed on the calculation result, resulting in a strange result as compared with Figure 2.3(c).

We can avoid this problem by a technique called **zero padding**, which gives the same result as the linear convolution. In general, if $u_1(m_1)$ and $h(m_1)$ are sampled by N and L points, respectively, expanding them to $N+L-1$ points eliminates the influence of the wraparound (we extend $u_1(m_1)$ and $h(m_1)$ like Figure 2.4(c)).

When the convolution is performed with the extended $u_1(m_1)$ and $h(m_1)$, the wraparound does not overlap the convolution result, as shown in Figure 2.4(d). The same result as the linear convolution is obtained by extracting necessary parts from the convolution result. Applying the zero padding technique to the diffraction calculation will be shown in Section 2.6.3.

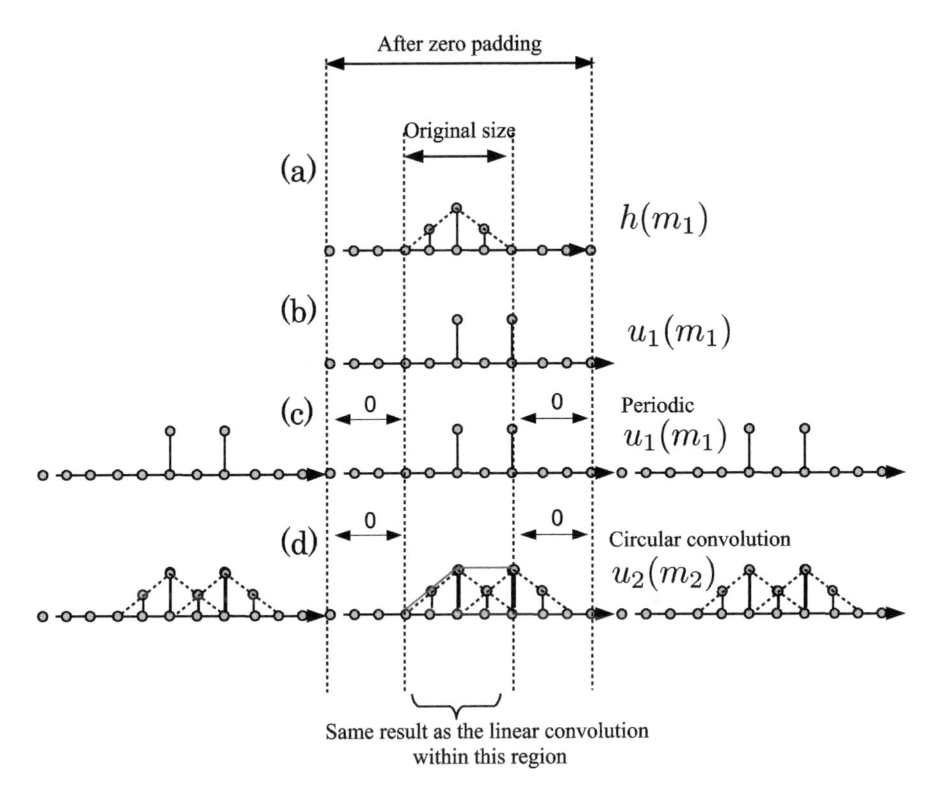

FIGURE 2.4 Converting circular convolution to linear convolution.

PERIODICITY OF TWO-DIMENSIONAL FFT

When handling a two-dimensional image with FFT, it is considered that the two-dimensional image is periodically expanded as shown in Figure 2.5. Since the two-dimensional convolution using FFT also causes wraparounds, it should expand to $2N \times 2N$ when the two-dimensional image has $N \times N$ pixels, and the extended part is filled with zeros as shown in Figure 2.6.

LOW-FREQUENCY AND HIGH-FREQUENCY POSITION OF THE SPECTRUM OBTAINED BY FFT

The two-dimensional FFT yields the spectrum shown on the left in Figure 2.7. The position of the low- and the high-frequency components are shown

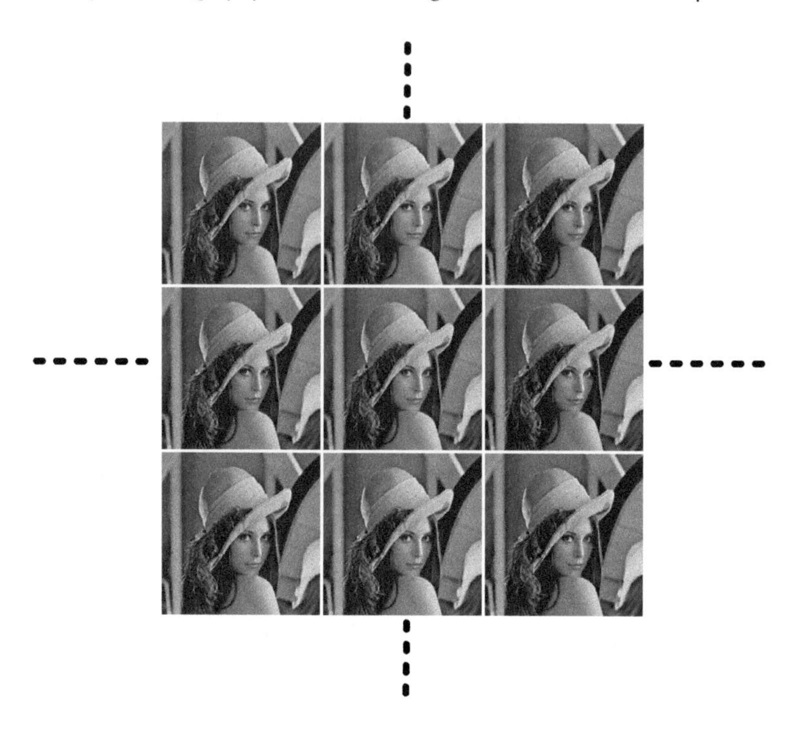

FIGURE 2.5 Periodicity of FFT.

in the left of Figure 2.7. The DC component is arranged at the position of the coordinate (0, 0) of the spectral image, and as it goes to the center of the spectral image, it becomes a high-frequency component.

We often need to exchange the first quadrant and the third quadrant of the spectral image and exchange the second quadrant and the fourth quadrant as indicated by the arrows. The exchanged spectrum shown on the right in Figure 2.7 is obtained. In this case, the center of the spectrum is the DC component and its frequency increases as it goes to the borders. The quadrant exchanging is called **quadrant exchange**, and is called **FFT shift** in standard programming environments 1such as MATLAB®, Octave, and Scilab.

Keeping these points in mind, let us implement the Fresnel diffraction of the convolution form using FFTs. The Fresnel diffraction using FFTs is represented by

$$u_2(m_2, n_2) = \mathrm{FFT}^{-1}\left[\mathrm{FFT}\left[u_1(m_1, n_1)\right]\mathrm{FFT}\left[\exp\left(\frac{i\pi p^2}{\lambda z}(m_1^2 + n_1^2)\right)\right]\right], \tag{2.49}$$

and it can be calculated according to the flow of Figure 2.8.

Although FFT may be implemented by itself, the efficient implementation is difficult. Currently, open source high-performance libraries of FFTs, such

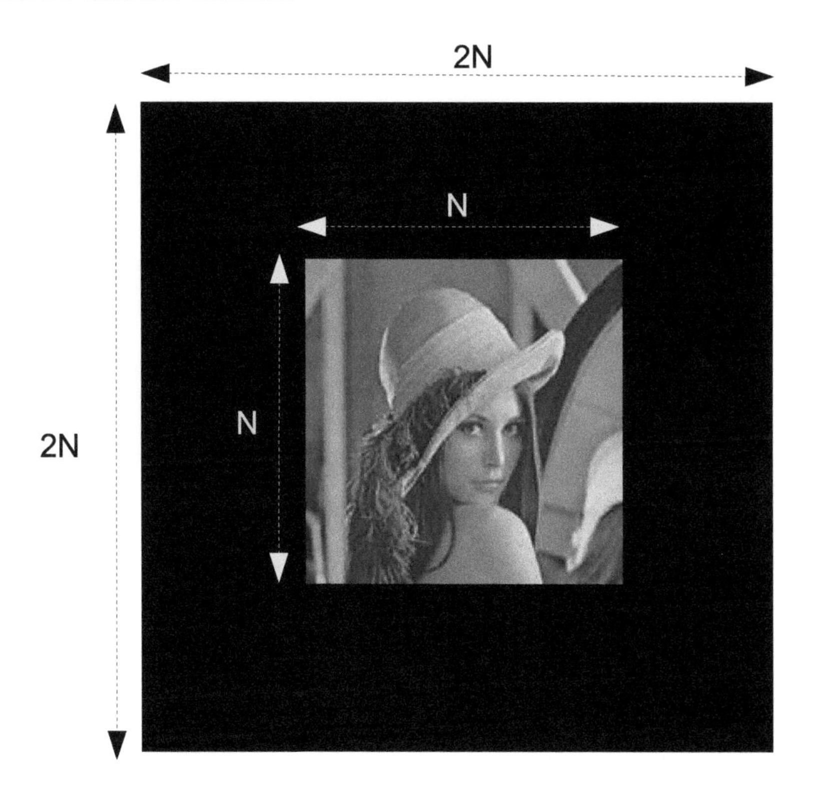

FIGURE 2.6 Zero padding.

as **FFTW**[vii] [17], have been developed. This book uses FFTW.

The impulse response $h(m_1, n_1) = \exp(\frac{i\pi p^2}{\lambda z}(m_1^2 + n_1^2))$ of Eq. (2.49) can be calculated using Listing 2.5.

Listing 2.5 Impulse response in Eq. (2.49).

```
1 void response(fftw_complex *h, int N, double lambda, double z,
      double p)
2 {
3   fftw_complex tmp;
4   tmp[0] = 0.0;
5   tmp[1] = 0.0;
6   for (int n=0; n<N; n++){
7     for (int m = 0; m<N; m++){
8       int idx = m + n*N;
```

[vii]Fastest Fourier Transform in the West. FFTW is a high-performance FFT library developed by Massachusetts Institute of Technology (MIT), which supports multithreading and vector operations, and is performed in various computing environments. Other well-known FFT libraries often have FFTW-compatible interfaces.

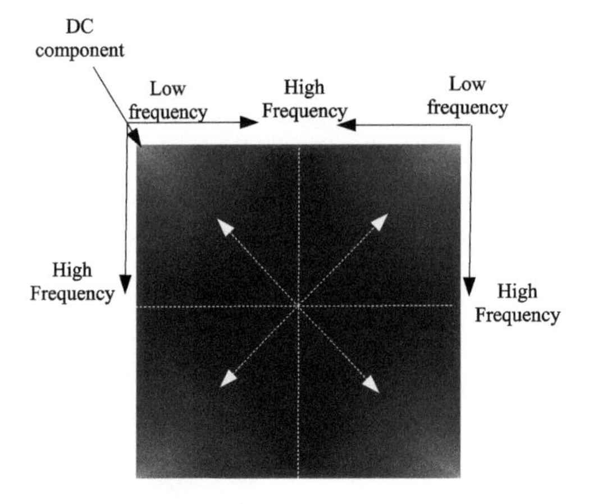

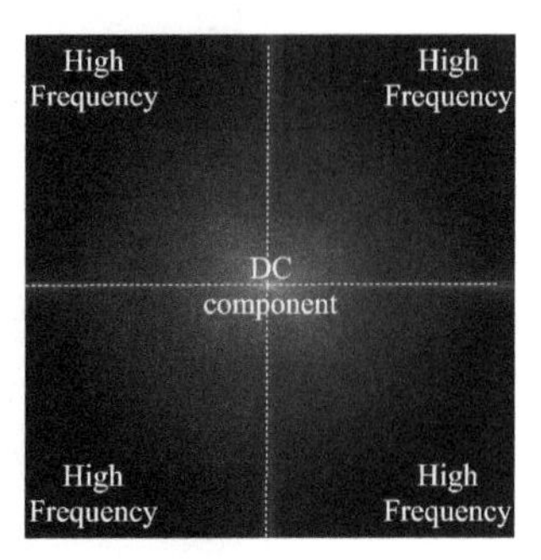

FIGURE 2.7 Spectrum obtained by FFT.

```
 9            double dx = (m - N / 2)*p;
10            double dy = (n - N / 2)*p;
11            double phase = (dx*dx + dy*dy)*M_PI / (lambda*z);
12            h[idx][0] = cos(phase);
13            h[idx][1] = sin(phase);
14        }
15    }
16 }
```

The calculation result of this function is stored in h indicated by the pointer of fftw_complex type. We can perform FFT and inverse FFT using the FFTW library as shown in Listing 2.6.

Listing 2.6 FFT and inverse FFT.

```
 1 void fft(fftw_complex *u1, fftw_complex *u2, int N)
 2 {
 3   fftw_plan plan = fftw_plan_dft_2d(
 4     N, N, u1, u2, FFTW_FORWARD, FFTW_ESTIMATE);
 5   fftw_execute(plan);
 6   fftw_destroy_plan(plan);
 7 }
 8
 9 void ifft(fftw_complex *u1, fftw_complex *u2, int N)
10 {
11   fftw_plan plan = (fftw_plan)fftw_plan_dft_2d(
12     N, N, u1, u2, FFTW_BACKWARD, FFTW_ESTIMATE);
13   fftw_execute(plan);
14   fftw_destroy_plan(plan);
15 }
```

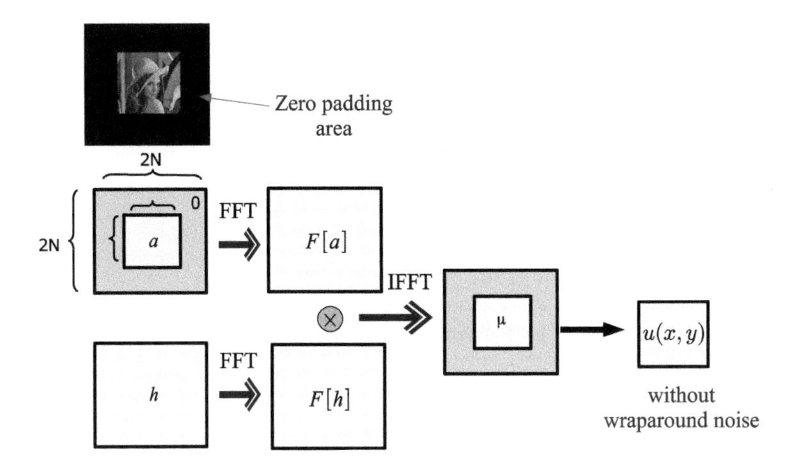

FIGURE 2.8 Fresnel diffraction in the convolution form by FFT.

FFTW is used as follows: First, we create a plan specifying the FFT size using function fftw_plan_dft_2d. Next, we execute FFT according to the plan using function fftw_execute. Finally, we destroy the plan using function fftw_destroy_plan. We can switch FFT and inverse FFT with FFTW_FORWARD and FFTW_BACKWARD.

According to Figure 2.8, implementing the Fresnel diffraction of the convolution form can be written as shown in Listing 2.7. The function fft_shift is the quadrant exchange. The function mul_complex is the complex multiplication, and the function mul_dbl is the scalar multiplication.

Listing 2.7 Fresnel diffraction of the convolution form using FFT.

```
1  void fft_shift(fftw_complex *u, int N)
2  {
3    int Nh = N / 2;
4    for (int i = 0; i<Nh; i++) {
5      for (int j = 0; j<Nh; j++) {
6
7        fftw_complex tmp1, tmp2;
8        int adr1, adr2;
9
10       adr1 = j + i*N;
11       adr2 = (j + Nh) + (i + Nh)*N;
12       tmp1[0] = u[adr1][0]; tmp1[1] = u[adr1][1];
13       tmp2[0] = u[adr2][0]; tmp2[1] = u[adr2][1];
14       u[adr1][0] = tmp2[0]; u[adr1][1] = tmp2[1];
15       u[adr2][0] = tmp1[0]; u[adr2][1] = tmp1[1];
16
17       adr1 = (j + Nh) + i*N;
```

```
18          adr2 = (j)+(i + Nh)*N;
19          tmp1[0] = u[adr1][0]; tmp1[1] = u[adr1][1];
20          tmp2[0] = u[adr2][0]; tmp2[1] = u[adr2][1];
21
22          u[adr1][0] = tmp2[0]; u[adr1][1] = tmp2[1];
23          u[adr2][0] = tmp1[0]; u[adr2][1] = tmp1[1];
24        }
25      }
26    }
27
28    void mul_complex(fftw_complex *a, fftw_complex *b, fftw_complex *c,
          int N)
29    {
30      fftw_complex tmp;
31      for (int i = 0; i < N*N; i++){
32        tmp[0] = a[i][0] * b[i][0] - a[i][1] * b[i][1];
33        tmp[1] = a[i][0] * b[i][1] + a[i][1] * b[i][0];
34        c[i][0] = tmp[0];
35        c[i][1] = tmp[1];
36      }
37    }
38
39    void mul_dbl(fftw_complex *u, double a, int N)
40    {
41      for (int i = 0; i<N*N; i++){
42        u[i][0] *= a;
43        u[i][1] *= a;
44      }
45    }
46
47    void fresnel_fft(fftw_complex *u1, int N, double lambda, double z,
          double p)
48    {
49      fft_shift(u1, N);
50      fft(u1, u1, N);
51
52      fftw_complex *u2 = (fftw_complex*)malloc(sizeof(fftw_complex)*N*
            N);
53      response(u2, N, lambda, z, p);
54      fft_shift(u2, N);
55      fft(u2, u2, N);
56
57      mul_complex(u1, u2, u1, N);
58      ifft(u1, u1, N);
59      fft_shift(u1, N);
60
61      mul_dbl(u1, 1.0 / (N*N), N);
62
```

```
63   free(u2);
64 }
```

2.6.3 EXAMPLES OF USE OF DIFFRACTION CALCULATION

Let us calculate the situation in Figure 2.9 using the Fresnel diffraction of Listing 2.7, in which the position of the diffraction pattern is 1 m away from an aperture and the aperture is illuminated by a plane wave with the wavelength $\lambda = 633$ nm (corresponding to red laser).

The source and the destination planes have $1,024 \times 1,024$ pixels, and the sampling interval is 10 μm. The rectangular aperture in the source plane has 50×50 pixels (physically, it is a hole with 500 μm $\times$ 500 μm) and the other area shields the incident light. The diffraction intensity pattern calculated by Listing 2.7 is shown on the right side of Figure 2.9[viii]. This simulation result agrees well with an optical experiment.

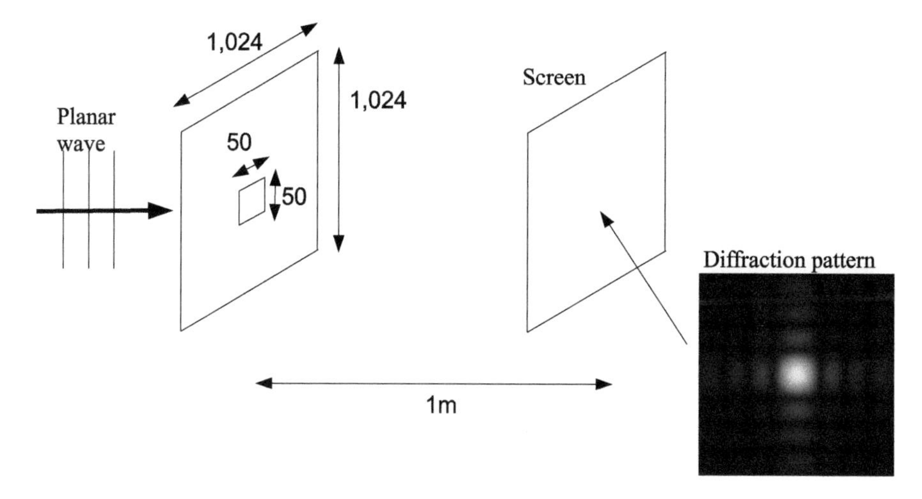

FIGURE 2.9 Simulation of rectangular aperture.

Let us see the effect of zero padding in the calculation result. Figure 2.10 shows the diffraction intensity pattern and the phase pattern when the aperture is placed at the center of the source plane (Figure 2.10(a)). As shown in Figure 2.10(b), the influence of the wraparound can be confirmed at the end of the diffracted intensity pattern without zero padding. The influence of the wraparound is more conspicuous in the phase pattern. This is interpreted as follows: if the diffracted light emitted from the aperture exceeds the calculation window as shown in Figure 2.10(c), the diffracted light of the exceeded part will wrap around from the opposite side.

[viii]Since the result of the Fresnel diffraction is the complex amplitude, we actually need to calculate the light intensity from the obtained complex amplitude.

On the other hand, the diffracted intensity pattern using zero padding has no influence on the wraparound. This can be interpreted as follows: if the diffracted light emitted from the aperture exceeds the calculation window, the diffracted light only affects the zero padding region, so it does not affect the calculation window.

Figure 2.11 shows the results when placing the rectangular aperture at the right end (Figure 2.11(a)). The effect of the zero padding is more pronounced. As can be seen from Figure 2.11(b) and (c), without using zero padding, the diffracted light emitted from the aperture wraps around from the left side, and it is confirmed that the wraparound affects both the intensity and phase patterns. On the other hand, the diffracted light with zero padding has no influence on the wraparound. When implementing the FFT-based diffraction calculation of the convolution form, zero padding is necessary.

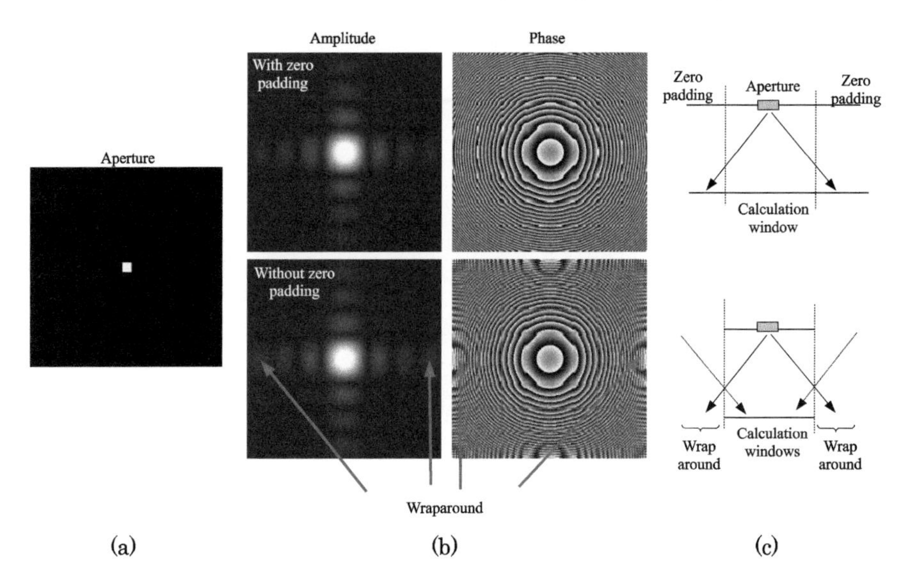

FIGURE 2.10 Diffraction results with/without zero padding.

Using the Fresnel diffraction can also simulate a lens imaging as shown in Figure 2.12. By **geometric optics**, the relationship between the object distance a and the image distance b can be shown to be

$$\frac{1}{a} + \frac{1}{b} = \frac{1}{f}, \tag{2.50}$$

where f is the focal length of the lens.[ix]

[ix]Geometric optics refers to a method of treating light as a ray without considering the wave nature of light. Since diffraction calculation deals with the wave nature of light, optical calculation using diffraction calculation is called **wave optics**.

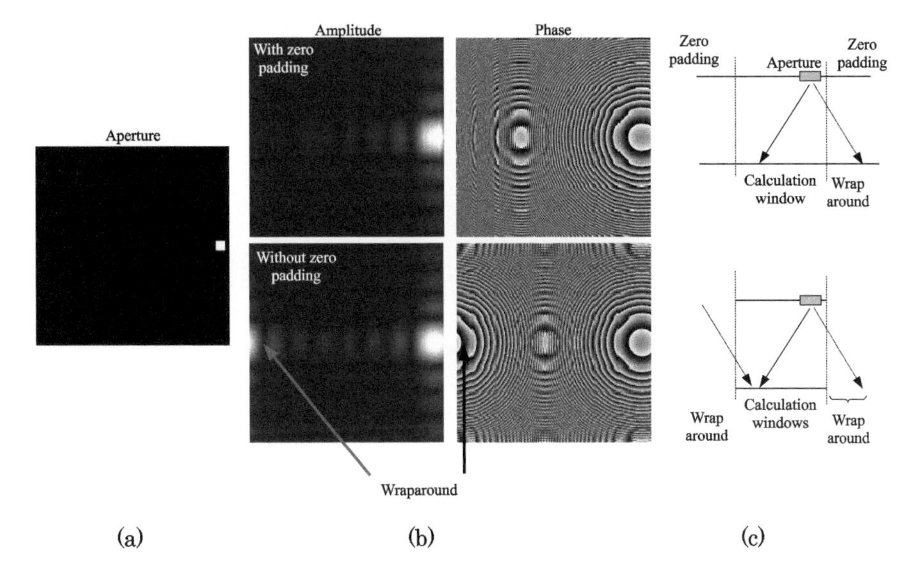

FIGURE 2.11 Diffraction results with/without zero padding.

The magnification (lateral magnification) M of the image formed by the lens is $M = b/a$. For example, if we use $a = 10$ cm, $b = 10$ cm and a lens with $f = 5$ cm, we will obtain the image with the lateral magnification of 1. Note that the image becomes an inverted image.

To simulate this lens, a plane wave is illuminated to the image and the Fresnel diffraction from the image to the lens is calculated. Subsequently, the calculation result is multiplied by the phase conversion of the lens,[x] and we calculate the Fresnel diffraction from the lens to the screen. The phase conversion of the lens with focal length f can be written as

$$\exp\left(-i\frac{2\pi}{\lambda}\left(\frac{x^2+y^2}{f}\right)\right), \tag{2.51}$$

using Fresnel approximation.

Since this expression is the same as the impulse response of the Fresnel diffraction (Eq. (2.43)) with $z = -f$, the source code in Listing 2.5 can be used. Listing 2.8 shows the source code of the lens simulation using the Fresnel diffraction. The calculation conditions are $a = 10$ cm, $b = 10$ cm, $f = 5$ cm and $\lambda = 633$ nm for the wavelength. The lateral magnification of the image is 1. We can obtain the inverted image as shown on the right in Figure 2.12.

[x]Lenses converge or diverge light by changing the phase of incident light. For example, the phase conversion of a convex lens focuses the incident light of a plane wave at the focal position; therefore, the phase conversion can be expressed by a spherical wave (Eq. (2.51)).

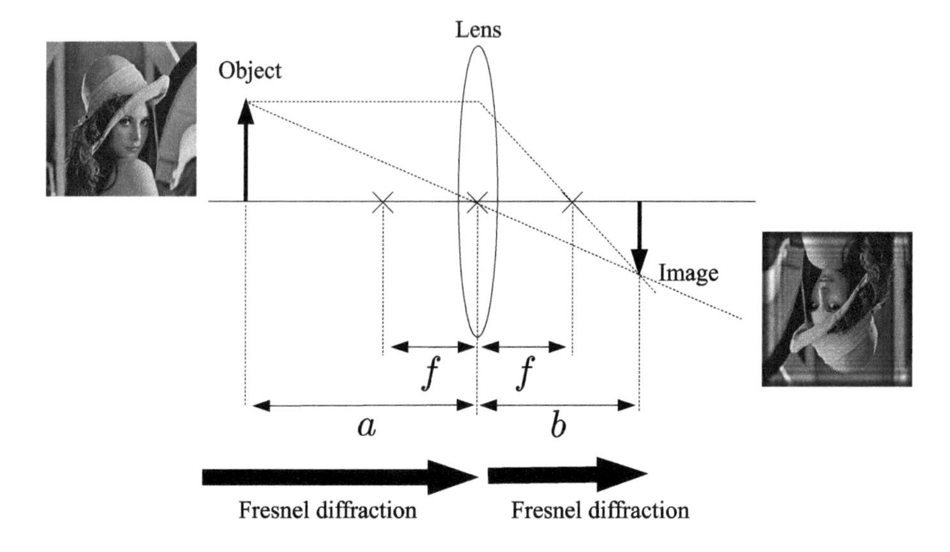

FIGURE 2.12 Simulation of lens.

Listing 2.8 Simulation of lens.

```
1 fresnel_fft(u, N, 633e−9, 10e−2, 10e−6);
2 response(u_lens, N, 633e−9, −5e−2, 10e−6);
3 mul_complex(u, u_lens, u, N);
4 fresnel_fft(u, N, 633e−9, 10e−2, 10e−6);
```

2.6.4 ALIASING IN DIFFRACTION CALCULATION

In numerical diffraction calculation, it is necessary to pay attention to **aliasing** that is caused by discretization. When discretizing a signal, according to the sampling theorem, aliasing occurs if the spatial variation of the signal is more than twice as fast as the sampling interval. In other words, it can be said that aliasing does not occur when the signal is discretized at a sampling frequency more than twice the maximum frequency of the signal.

Let us consider the Fresnel diffraction of the convolution form (Eq. (2.24)). We rewrite the equation as

$$u_2(x_2, y_2) = \frac{\exp(i\frac{2\pi}{\lambda}z)}{i\lambda z} \times \mathcal{F}^{-1}\Big[\mathcal{F}\Big[u_1(x_1, y_1)\Big] \cdot \mathcal{F}\Big[h(x_1, y_1)\Big]\Big], \tag{2.52}$$

where $h(x, y)$ is defined as

$$h(x, y) = \exp\left(i\frac{\pi}{\lambda z}(x^2 + y^2)\right). \tag{2.53}$$

In Eq. (2.52), the spatially varying signal is $h(x, y)$, which is the cause of aliasing. Figure 2.13 illustrates the real part of $h(x, y)$ when the propagation distance z is $z = 0.26$ m (Figure 2.13(a)) and $z = 0.05$ m (Figure 2.13(b)). The calculation conditions are that the sampling interval is $p = 10$ μm, wavelength λ =633 nm, number of pixels $N \times N = 512 \times 512$.

The graphs show the change on the x axis of $h(x, y)$. In the left graph, the sampling is performed at an appropriate sampling interval, so that aliasing does not occur. In the right graph, aliasing occurs because the sampling interval is coarse for the change of $h(x, y)$.

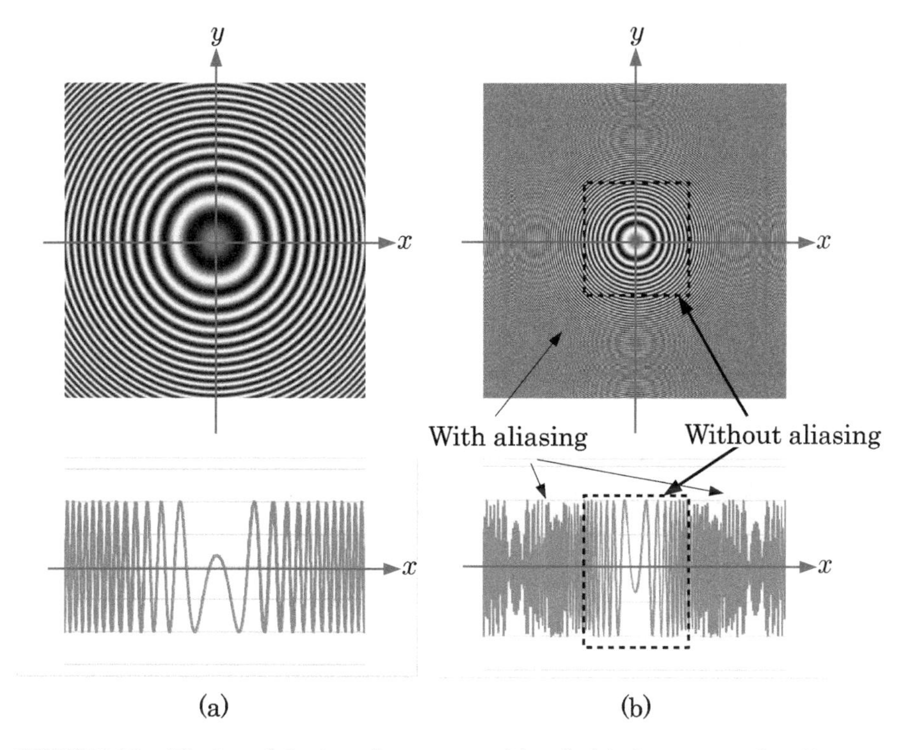

FIGURE 2.13 Aliasing of the impulse response $h(x, y)$: (a) the propagation distance z is $z = 0.26$ m and (b) $z = 0.05$ m.

The condition under which aliasing occurs is determined from the **spatial frequency** of $h(x, y)$. For example, when a one-dimensional signal is expressed as

$$\exp(2\pi i f x) = \exp(i\phi(x)), \tag{2.54}$$

where $\phi(x) = 2\pi f x$ and f is the frequency, the frequency can be obtained by

$$\frac{1}{2\pi}\frac{d\phi(x)}{dx} = f. \tag{2.55}$$

Similarly, when a two-dimensional signal is expressed as

$$\exp(2\pi i(f_x x + f_y y)) = \exp(i\phi(x, y)), \tag{2.56}$$

where $\phi(x, y) = 2\pi(f_x x + f_y y)$ with the horizontal and vertical spatial frequencies f_x and f_y, these spatial frequencies are calculated by

$$\frac{1}{2\pi}\frac{\partial\phi(x, y)}{\partial x} = f_x, \quad \frac{1}{2\pi}\frac{\partial\phi(x, y)}{\partial y} = f_y. \tag{2.57}$$

Defining $\phi(x, y) = \frac{\pi}{\lambda z}(x^2 + y^2)$ in Eq. (2.53), the spatial frequencies f_x and f_y of $h(x, y)$ are obtained as

$$f_x = \frac{1}{2\pi}\frac{\partial\phi(x, y)}{\partial x} = \frac{x}{\lambda z}, \quad f_y = \frac{1}{2\pi}\frac{\partial\phi(x, y)}{\partial y} = \frac{y}{\lambda z}. \tag{2.58}$$

Since these frequencies vary with position, it is also called **local spatial frequency**. We can see that the spatial frequency of $h(x, y)$ is proportional to (x, y). Such a signal is called a **chirp signal**.

From Eq. (2.58) and Figure 2.13, the maximum spatial frequency is the edge of $h(x, y)$ ($x_{max} = \frac{N}{2}p$) where p is the sampling interval. In order not to cause aliasing, it is necessary to sample the signal at a frequency more than twice the maximum frequency.[xi] Therefore, the following condition should be satisfied:

$$\begin{aligned} \frac{1}{p} &\geq 2\left|\frac{1}{2\pi}\frac{\partial\phi(x, y)}{\partial x}\right| = \left|\frac{2x}{\lambda z}\right|, \\ \frac{1}{p} &\geq 2\left|\frac{1}{2\pi}\frac{\partial\phi(x, y)}{\partial y}\right| = \left|\frac{2y}{\lambda z}\right|. \end{aligned} \tag{2.59}$$

The z that does not cause aliasing will be

$$z \geq \frac{Np^2}{\lambda} \tag{2.60}$$

by substituting $x_{max} = \frac{N}{2}p$ (the edge of $h(x, y)$) into Eq. (2.59). In the calculation condition of Figure 2.13(a), aliasing does not occur because it is roughly $z \geq 8$ cm. In contrast, aliasing occurs in the calculation condition of Figure 2.13(b).

Similarly, find the lateral range that does not cause aliasing (dashed square in Figure 2.13(b)). Solving Eq. (2.59) for x, y yields

$$|x| \leq \frac{\lambda z}{2p}, \tag{2.61}$$

$$|y| \leq \frac{\lambda z}{2p}. \tag{2.62}$$

[xi] If x, y is negative, it becomes negative spatial frequency. Therefore, it takes the absolute value.

In the condition of Figure 2.13(b), aliasing is not affected within about 320 pixels from the center in pixel units, but the outer areas are affected by aliasing.

Figure 2.14(a) shows the calculation result with $z = 0.26$ m using the Fresnel diffraction in the convolution form under the same condition as Figure 2.13. Aliasing does not occur. Figure 2.14(a) shows the calculation result with $z = 0.05$ m using the Fresnel diffraction in the convolution form. Figure 2.14(b) causes intense aliasing.

In addition, Figure 2.14(c) shows the calculation result with $z = 0.05$m using the angular spectrum method under the same condition as Figure 2.13(b). In the angular spectral method, the clean diffracted result can be obtained without aliasing despite the fact that it is a short propagation distance ($z = 0.05$ m). Contrary to the Fresnel diffraction of the convolution form, although the angular spectrum method can perform accurate diffraction calculation without approximation, it is known that aliasing occurs at a long distance propagation.

This reason can be explained from the impulse response of the Fresnel diffraction (Eq. (2.53)) and the transfer function of the angular spectral method (Eq. (2.10)).

Since the impulse response of the Fresnel diffraction is

$$h(x, y) = \exp(i\frac{\pi}{\lambda z}(x^2 + y^2)). \tag{2.63}$$

Because z is in the denominator, the spatial frequency of $h(x, y)$ changes gently as z increases. Therefore, aliasing becomes less likely to occur.

Conversely, the transfer function of the angular spectral method is

$$H(f_x, f_y) = \exp\left(i2\pi z\sqrt{\frac{1}{\lambda^2} - f_x^2 - f_y^2}\right). \tag{2.64}$$

Because z is in the numerator, the spatial frequency of $H(f_x, f_y)$ violently vibrates as z increases, and aliasing tends to occur. When z is small, the spatial frequency of $H(f_x, f_y)$ changes gently.

From the viewpoint of aliasing, the Fresnel diffraction in the convolution form is suitable for a long propagation, and the angular spectrum method is suitable for a short propagation.

2.7 APODIZATION OF RINGING ARTIFACTS

Another problem of diffraction calculations is **ringing artifacts**. Figure 2.15(a) shows the result of a diffraction calculation with the distance of $z = 0.1$ m, the sampling interval of 10 μm and the wavelength of 633 nm. The ringing artifact is the vibration appearing around the diffraction result. This is caused by abrupt change in the luminance value at the boundary of the calculation

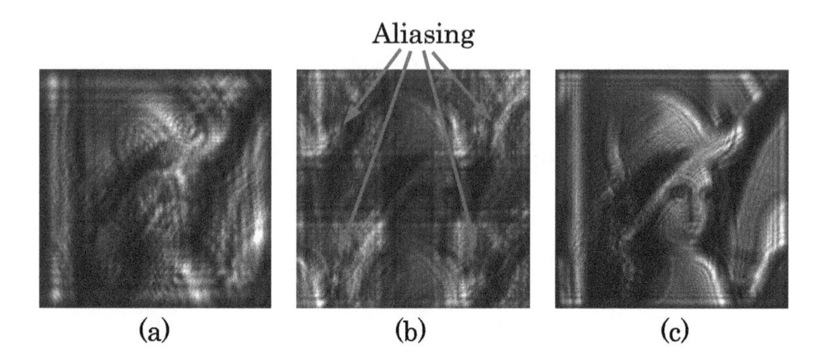

FIGURE 2.14 Comparison of Fresnel diffraction and angular spectrum method.

window.[xii] The ringing artifact can be suppressed by applying a window function, which smooths the change in boundary brightness, to the calculation window [18–21]. Another method has been proposed in which the calculation window is expanded to reduce the change in boundary brightness [22]. Such methods to mitigate the ringing artifacts are referred as to **apodization.**

Here, we introduce apodization using window functions. Window functions are often used in the field of signal processing. For example, Hanning and Hamming windows are often used. The one-dimensional **Hamming window** is defined by

$$w(m) = 0.5\left(1 - \cos 2\pi \frac{m}{M-1}\right). \tag{2.65}$$

When extending a one-dimensional window function $w_1(m)$ to the two-dimensional window function $w_1(m,n)$, there are two expressions. The separate expression is defined as

$$w_2(m,n) = w_1(m)w_1(n). \tag{2.66}$$

The symmetric expression is defined as

$$w_2(m,n) = w_1(\sqrt{m^2+n^2}). \tag{2.67}$$

We perform apodized diffraction calculation using the window function $w_2(m,n)$ by

$$u_2(m_2,n_2) = \text{Prop}_z[u_1(m_1,n_1)w_2(m,n)]. \tag{2.68}$$

As shown in Figure 2.15(b), although the ringing artifacts can be reduced, there is a disadvantage that the periphery of the diffraction pattern becomes dark because this window function has few flat areas.

[xii]In the Fourier series theory, it is well known as the **Gibbs phenomenon**.

One of the flat-headed window functions has a **Tukey window** [20] defined as

$$w(m) = \begin{cases} 0.5\left(1+\cos(\frac{2\pi}{r}\frac{m-1}{M-1}-\pi)\right) & m < \frac{r}{2}(M-1), \\ 0.5\left(1+\cos(\frac{2\pi}{r}(1-\frac{m-1}{M-1})-\pi)\right) & m > (M-1)-\frac{r}{2}(M-1), \\ 1.0 & otherwise. \end{cases} \tag{2.69}$$

In addition, other flat-headed $\cos^k$ window function has been proposed [19]. The window function is defined as

$$w(m) = \begin{cases} \cos^k\left(a\left(\frac{2(m+1)}{L-L_0}-1\right)\right) & m+1 \leq \frac{L-L_0}{2}, \\ 1 & \frac{L-L_0}{2} < m+1 \leq \frac{L+L_0}{2}, \\ \cos^k\left(b\left(\frac{2(m+1)}{L+L_0+2}-1\right)\right) & otherwise, \end{cases} \tag{2.70}$$

where $a = -\pi\frac{L-L_0}{L_0+2-L}$ and $b = \pi\frac{L+L_0+2}{L_0-L_0-2}$.

Figure 2.15(c) shows the diffraction pattern when using the Tukey window. Figure 2.15(d) shows the diffraction pattern using the flat-headed $\cos^k$ window function. Compared with Figure 2.15(b), these the flat-headed window functions confirm that the ringing artifacts can be reduced while maintaining the area of the calculation window.

2.8 SPECIAL DIFFRACTION CALCULATIONS

As in Figure 2.16, the limitations of the Fresnel diffraction and angular spectral method that are often used in computer holography are as follows:

- The source plane u_1 and the destination plane u_2 must be coaxial.
- The sampling intervals of the source plane u_1 and destination plane u_2 cannot be freely set.
- The source plane u_1 and the destination plane u_2 must be parallel.

In this section, we introduce a **shift diffraction calculation** that can be used even when the source plane u_1 and the destination plane u_2 are off-axis as shown in Figure 2.17(a). We also describe a **scale diffraction calculation** that allows us to freely set the sampling intervals between the source plane u_1 and the destination plane u_2.

In Section 3.3, we will explain **tilt diffraction calculation**, which can be calculated even if the source and destination planes are nonparallel, as shown in Figure 2.17(b). Such diffraction calculation is useful, for example, when calculating a hologram from a three-dimensional object composed of polygons, since each polygon is not parallel to the hologram.

2.8.1 SHIFT DIFFRACTION AND SCALE DIFFRACTION

Various schemes for shift diffraction calculation and scale diffraction calculation have been proposed [23–31]. Here, we introduce the Fresnel diffraction

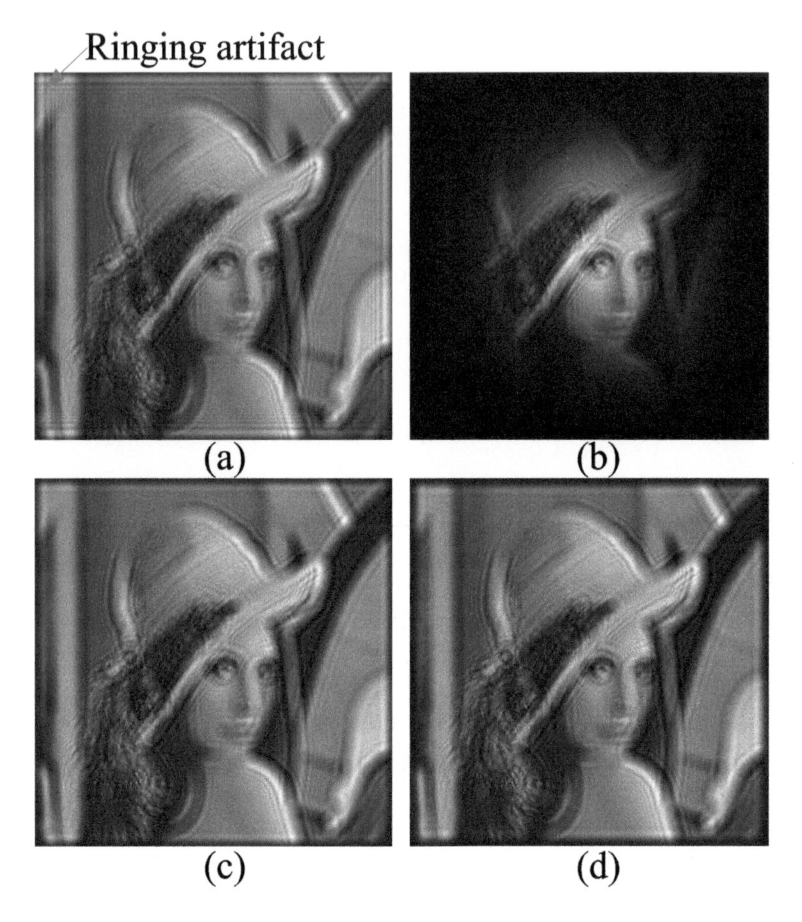

FIGURE 2.15 Comparison of diffraction results using window functions: (a) original diffraction pattern, (b) Hamming window, (c) Tukey window, and (d) $\cos^k$ window function.

that can be performed together with the shift and scale diffraction calculation proposed in [24], but the derivation of the formula follows the other literature [29].[xiii]

Let us discretize the destination plane $u_2(x_2, y_2)$ with the sampling interval p as shown in Figure 2.18. The sampling interval of the source plane $u_1(x_1, y_1)$ is defined by multiplying p with a rate of s. When $s = 1$, it is a normal Fresnel diffraction. When $s > 1$, the source plane becomes larger than the destination plane. When $s < 1$, the area of the source decreases.

[xiii]Scale and shift diffraction calculation by [29] can suppress aliasing occurring in the original method [24].

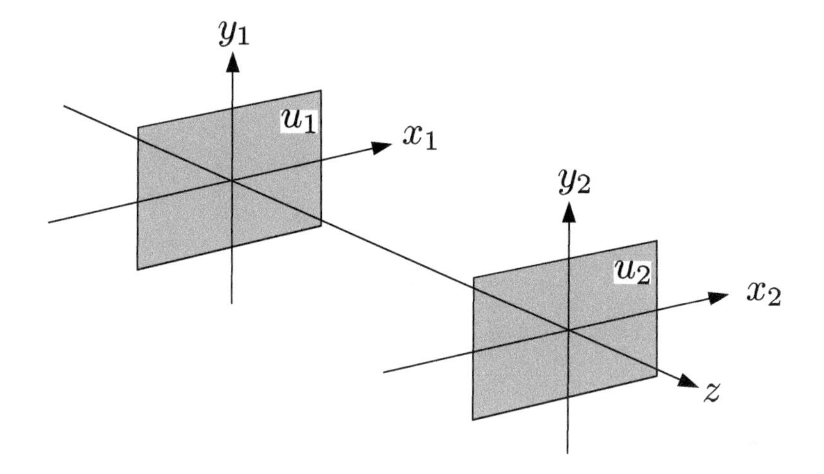

FIGURE 2.16 Limitations of the Fresnel diffraction and angular spectral method.

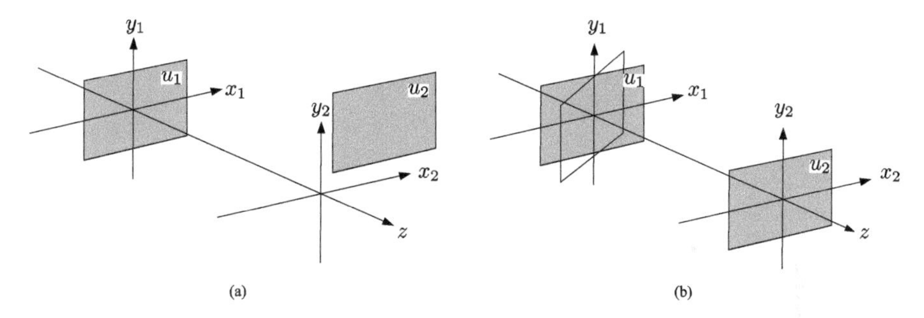

FIGURE 2.17 Special diffraction calculations: (a) shift diffraction calculation and scale diffraction calculation and (b) tilt diffraction calculation.

The Fresnel diffraction described in Section 2.3 is rewritten as follows:

$$
\begin{aligned}
u_2(x_2, y_2) &= \frac{\exp(ikz)}{i\lambda z} \int\int u_1(x_1, y_2) \\
&\quad \times \exp\left(\frac{i\pi}{\lambda z}((x_2 - x_1)^2 + (y_2 - y_1)^2)\right) dx_1 dx_2. \quad (2.71)
\end{aligned}
$$

As in Figure 2.18, for performing the scale diffraction, we multiply the coordinate (x_1, y_1) with s. For performing the shift diffraction, the destination u_2 is located away from the optical axis by (o_x, o_y). $(x_2 - x_1)^2$ and $(y_2 - y_1)^2$ in Eq. (2.71) is expressed as

$$
\begin{aligned}
(x_2 - sx_1 + o_x)^2 &= s(x_2 - x_1)^2 + (s^2 - s)x_1^2 + (1 - s)x_2^2 + 2o_x x_2 \\
&\quad -2so_x x_1 + o_x^2, \\
(y_2 - sy_1 + o_y)^2 &= s(y_2 - y_1)^2 + (s^2 - s)y_1^2 + (1 - s)y_2^2 + 2o_y y_2 \\
&\quad -2so_y y_1 + o_y^2. \quad (2.72)
\end{aligned}
$$

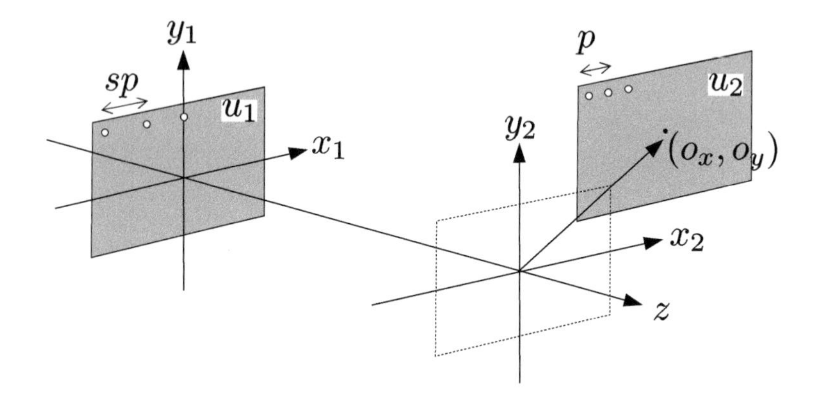

FIGURE 2.18 Scale and shift diffraction calculation.

Substituting Eq. (2.72) into Eq. (2.71), we can obtain

$$\begin{aligned} u_2(x_2, y_2) &= C_z \int\int u_1(x_1, y_1) \\ &\times \exp\left(\frac{i\pi}{\lambda z}((s^2 - s)x_1^2 - 2so_x x_1) + (s^2 - s)y_1^2 - 2so_y y_1)\right) \\ &\times \exp\left(\frac{i\pi(s(x_2 - x_1)^2 + s(y_2 - y_1)^2)}{\lambda z}\right) dx_1 dy_1. \end{aligned} \quad (2.73)$$

This equation is a convolution; therefore, we can rewrite this equation using the convolution theorem as follows:

$$u_2(x_2, y_2) = C_z \mathcal{F}^{-1}\left[\mathcal{F}\left[u_1(x_1, y_1)\exp(i\phi_u)\right]\mathcal{F}\left[\exp(i\phi_h)\right]\right], \quad (2.74)$$

where we define $\exp(i\phi_u)$ and $\exp(i\phi_h)$, C_z as

$$\begin{aligned} \exp(i\phi_u) &= \exp\left(i\pi \frac{(s^2 - s)x_1^2 - 2so_x x_1}{\lambda z}\right) \\ &\times \exp\left(i\pi \frac{(s^2 - s)y_1^2 - 2so_y y_1}{\lambda z}\right), \end{aligned} \quad (2.75)$$

$$\exp(i\phi_h) = \exp\left(i\pi \frac{sx_1^2 + sy_1^2}{\lambda z}\right), \quad (2.76)$$

$$\begin{aligned} C_z &= \frac{\exp(i\phi_c)}{i\lambda z} = \frac{\exp(ikz)}{i\lambda z} \\ &\times \exp\left(\frac{i\pi}{\lambda z}((1 - s)x_2^2 + 2o_x x_2 + o_x^2)\right) \\ &\times \exp\left(\frac{i\pi}{\lambda z}((1 - s)y_2^2 + 2o_y y_2 + o_y^2)\right). \end{aligned} \quad (2.77)$$

When computing Eq. (2.74) in a computer, it is similar to the procedure of implementing the Fresnel diffraction described in Section 2.6.2. It also notes that the convolution calculation using FFTs becomes circular convolution; therefore, we need zero padding before calculating Eq. (2.74).

Figure 2.19 shows the results of the scale diffraction calculation using Eq. (2.74). The sampling interval of the destination plane is defined by p, and the sampling interval of the source plane is defined by sp. The calculation conditions are $p = 10\mu$m, the wavelength of 633 nm, and the propagation distance of 0.5 m.

The center of the figure shows the diffraction pattern in $s = 1$ (thus, normal Fresnel diffraction). The left of the figure shows the diffraction pattern in $s = 0.5$, thus the sampling interval of the destination plane is $sp = 5$ μm. The right of the figure shows the diffraction pattern in $s = 2.0$, thus the sampling interval of the destination plane is $sp = 20$ μm.

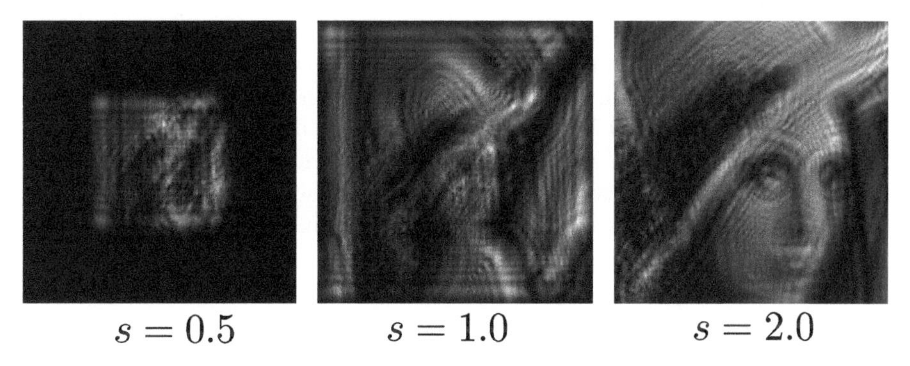

FIGURE 2.19 Diffraction patterns of the scale diffraction calculation using Eq. (2.74).

Figure 2.20 shows the diffraction patterns of the shift diffraction calculation using Eq. (2.74). In this calculation, we used the sampling intervals of 10 μm in the source and destination planes. The left figure shows the diffraction pattern when the offset (o_x, o_y)=(0 mm,0 mm), thus normal Fresnel diffraction. The center and right figures show the diffraction patterns when we use the offsets (o_x, o_y)=(5 mm,0 mm) and (o_x, o_y)=(5 mm,5 mm), respectively.

Finally, examples of the shift diffraction calculation and scale diffraction calculation are introduced. With the improvement of computer performance, opportunities for performing large-scale hologram generation and reconstructed images from large-scale holograms are increasing. Such calculations may require a large amount of memory that cannot be recorded in the memory of a computer at one time. The shift diffraction calculation is useful for that situation.

If the source plane u_1 and destination u_2 plane cannot be recorded in the memory of a computer at a time, we divide the calculation area into appropriate areas that can be recorded in the computer memory as shown in

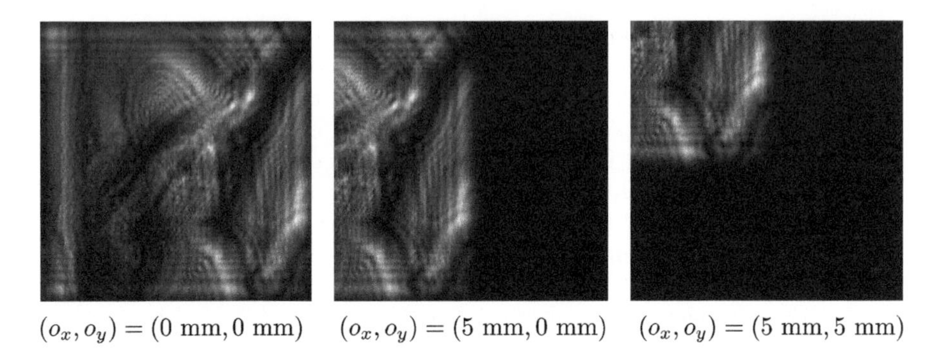

FIGURE 2.20 Diffraction patterns using shift diffraction with different offsets.

Figure 2.19 (this case has four divisions).

A diffracted light from the region A in the source plane is propagated to the regions A, B, C and D in the destination plane. Viewing from the region A in the source plane, the shift diffraction calculations are required for calculating the regions B, C and D because the optical axis of these regions does not coincide with the region A. A diffracted light from the remaining regions in the source plane can be calculated by the shift diffraction in the same manner, and the diffracted results are superimposed to obtain the whole diffracted field from the source plane.

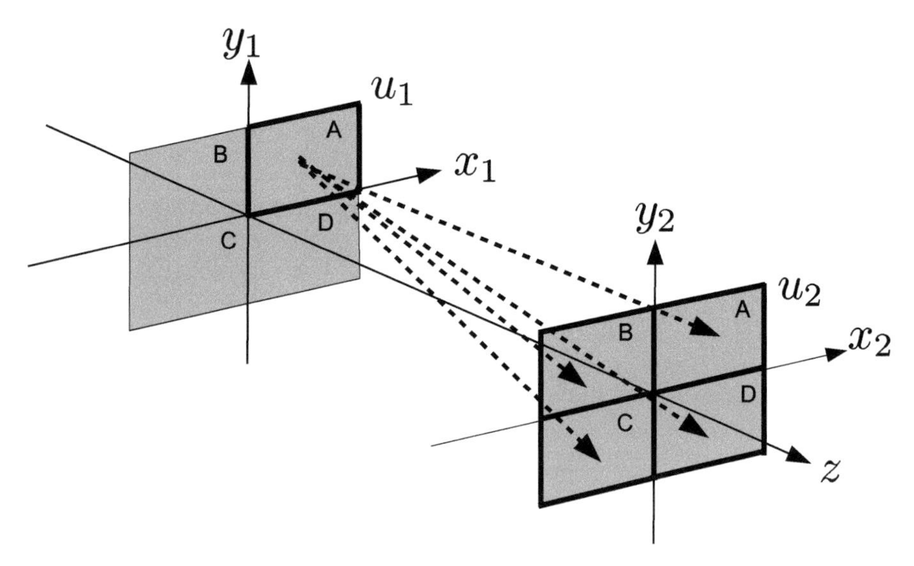

FIGURE 2.21 Diffraction calculation for a large area using shift diffraction.

The scale diffraction calculations have applications. For example, the applications are holographic projections that zoom the reconstructed image without using lenses, as described in Section 5.3.2. In addition, the application is to observe details of reconstructed images from an optically recorded hologram [28] that is captured by a digital holographic microscope.[xiv] Figure 2.22 shows the reconstructed image of a digital holographic microscope. In the upper figure, the whole reconstructed image is observed by normal diffraction calculation. In the bottom figure, the area indicated by the dashed box is enlarged 2.4 times by the scale diffraction calculation.

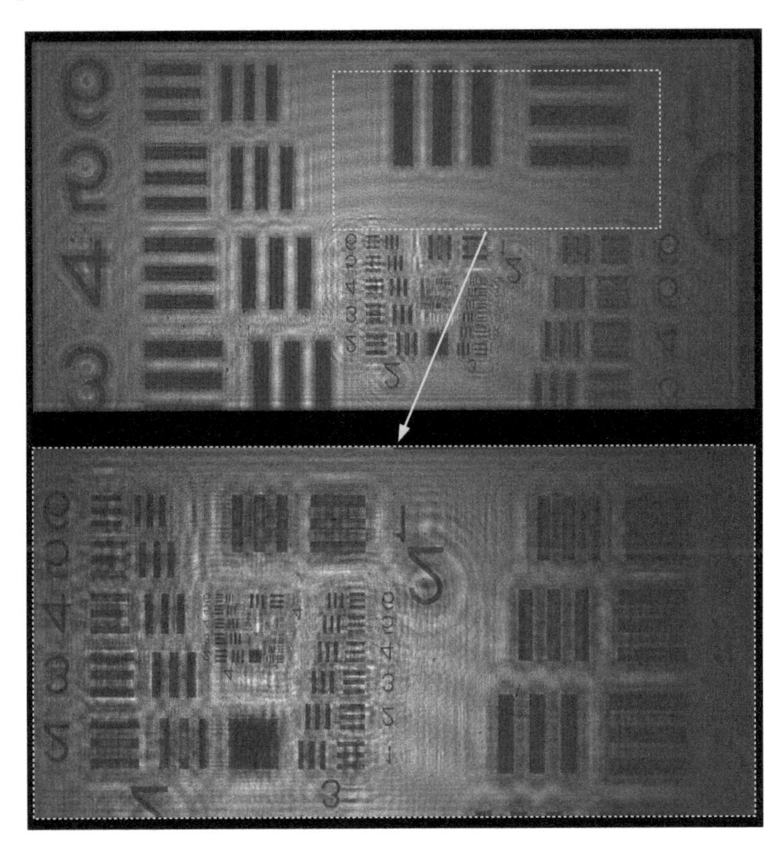

FIGURE 2.22 Changing magnification of digital holographic microscopy using the scale diffraction.

2.8.2 DOUBLE-STEP FRESNEL DIFFRACTION

We now introduce diffraction calculation with good calculation efficiency. In Fourier optics, diffraction calculations are categorized into two forms: the first

[xiv]The digital holographic microscope will be described in Chapter 4.

is convolution-based diffraction and the second is Fourier transform-based diffraction. The convolution-based diffraction is generally expressed as

$$u_2(x_2,y_2) = u_1(x_1,y_1) \otimes p_z(x_1,y_1) = \mathcal{F}^{-1}\Big[\mathcal{F}\Big[u_1(x_1,y_1)\Big]P_z(f_x,f_y)\Big], \quad (2.78)$$

where $\otimes$ denotes the convolution, the operators $\mathcal{F}\big[\cdot\big]$ and $\mathcal{F}^{-1}\big[\cdot\big]$ are the Fourier and inverse Fourier transforms, respectively, $u_1(x_1,y_1)$ and $u_2(x_2,y_2)$ indicate the source and destination planes, p_z is a point spread function, and $P_z(f_x,f_y) = \mathcal{F}\big[p_z(x_1,y_1)\big]$ is the transfer function. For example, the angular spectrum method uses $P_z(f_x,f_y) = \exp\left(-2\pi iz\sqrt{1/\lambda^2 - f_x^2 - f_y^2}\right)$.

An advantage of the convolution-based diffraction is that the sampling interval in the destination plane becomes the same as that in the source plane; however, a disadvantage is the need to expand the source and destination planes using zero padding to avoid the **aliasing** that occurs due to the circular convolution property of Eq.(2.78). It requires a large amount of memory and long calculation time.

Meanwhile, as described in Section 2.3.2, **single-step Fresnel diffraction** (SSF) is a Fourier transform-based diffraction. The SSF is expressed as

$$\begin{aligned} u_2(x_2,y_2) &= C_z \int\int u_1(x_1,y_1)\exp(\frac{i\pi(x_1^2+y_1^2)}{\lambda z}) \\ &\times \exp(\frac{-2\pi i(x_1x_2+y_1y_2)}{\lambda z})dx_1dy_1, \end{aligned} \quad (2.79)$$

where $C_z = \frac{\exp(ikz)}{i\lambda z}$.

The numerical version of the SSF (Eq.(2.82)) is obtained by defining the following:

$$\begin{aligned} (x_1,y_1) &= ((m_1 - N_x/2)p_{x_1},(n_1 - N_x/2)p_{y_1}), \\ (x_2,y_2) &= ((m_2 - N_x/2)p_{x_2},((n_2 - N_y/2)p_{y_2}), \end{aligned} \quad (2.80)$$

where $m_1, m_2 \in [0, N_x/2-1]$ and $n_1, n_2 \in [0, N_y/2-1]$, and the sampling rates on the source plane are p_{x_1} and p_{y_1}. The sampling rates on the destination plane are expressed as

$$\begin{aligned} p_{x_2} &= \lambda z/(N_x p_{x_1}), \\ p_{y_2} &= \lambda z/(N_y p_{y_1}). \end{aligned} \quad (2.81)$$

Eventually, the numerical version of the SSF is expressed as

$$\begin{aligned} u_2(m_2,n_2) &= \mathrm{SSF}_z\Big[u_1(m_1,n_1)\Big] \\ &= C_z\mathrm{FFT}\Big[u_1(m_1,n_1)\exp(\frac{i\pi(x_1^2+y_1^2)}{\lambda z})\Big], \end{aligned} \quad (2.82)$$

where the pixel numbers of the source and destination planes are $N_x \times N_y$.

The SSF can calculate the light propagation at z by a single FFT so that it does not need zero padding, unlike the convolution-based diffraction. Thus, it is an efficient approach in terms of the memory and the calculation time required; however, as shown in Eq. (2.81), the sampling rates on the destination plane are changed by the wavelength and propagation distance.

To overcome this problem, **double-step Fresnel diffraction** (DSF) was proposed [23]. As shown in Figure 2.23, it calculates the light propagation between the source plane and the destination plane by the two SSFs, via a virtual plane. The first SSF calculates the light propagation between the source plane and the virtual plane at distance z_1. The sampling rates on the virtual plane are

$$\begin{aligned} p_{x_v} &= \lambda z_1/(N_x p_{x_1}), \\ p_{y_v} &= \lambda z_1/(N_y p_{y_1}). \end{aligned} \tag{2.83}$$

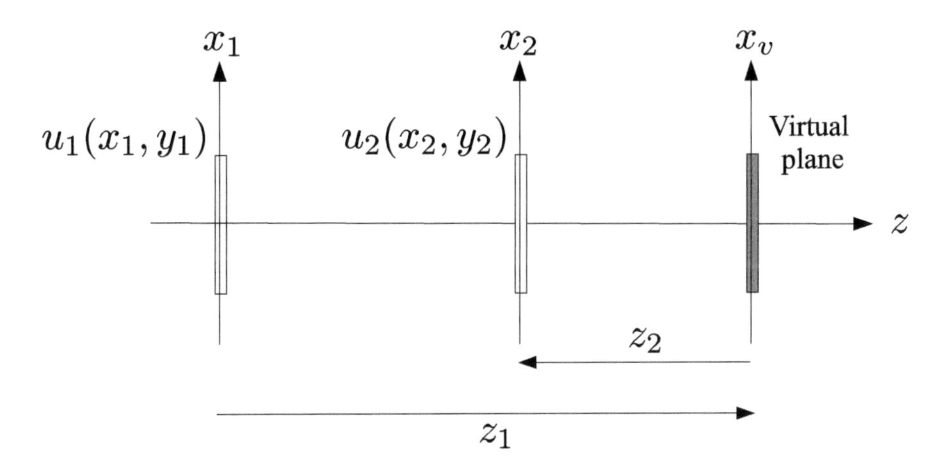

FIGURE 2.23 Double-step Fresnel diffraction.

The second SSF calculates the light propagation between the virtual plane and the destination plane at distance z_2. The sampling rates on the destination plane are

$$\begin{aligned} p_{x_2} &= \lambda z_2/(N_x p_{x_v}) = |z_2/z_1| p_{x_1}, \\ p_{y_2} &= \lambda z_2/(N_y p_{y_v}) = |z_2/z_1| p_{y_1}. \end{aligned} \tag{2.84}$$

The total propagation distance is $z = z_1 + z_2$ where z_1 and z_2 are acceptable for minus distance. The DSF is finally expressed as

$$\begin{aligned} u_2(m_2,n_2) &= \mathrm{DSF}_z\Big[u_1(m_1,n_1)\Big] = \mathrm{SSF}_{z_2}\Big[\mathrm{SSF}_{z_1}\Big[u_1(m_1,n_1)\Big]\Big] \\ &= C_{z_2}\mathrm{FFT}^{sgn(z_2)}[\exp(\frac{i\pi z(x_v^2+y_v^2)}{\lambda z_1 z_2}) \\ &\quad \times \mathrm{FFT}^{sgn(z_1)}[u_1(m_1,n_1)\exp(\frac{i\pi(x_1^2+y_1^2)}{\lambda z_1})]]. \end{aligned} \tag{2.85}$$

The operator $\mathrm{FFT}^{sgn(z)}$ means forward FFT when the sign of z is plus and inverse FFT when it is minus. In this equation, three terms of C_{z_2}, $\exp\left(\frac{i\pi z(x_v^2+y_v^2)}{\lambda z_1 z_2}\right)$, and $\exp\left(\frac{i\pi(x_1^2+y_1^2)}{\lambda z_1}\right)$ will cause aliasing.

The DSF introducing the rectangular function for band limitation, which is referred as to **band-limited double-step Fresnel diffraction** (BL-DSF) [32], is expressed as follows:

$$\begin{aligned} u_2(m_2,n_2) &= \mathrm{DSF}_z\Big[u_1(m_1,n_1)\Big] = \mathrm{SSF}_{z_2}\Big[\mathrm{SSF}_{z_1}\Big[u_1(m_1,n_1)\Big]\Big] \\ &= C_{z_2}\mathrm{FFT}^{sgn(z_2)}[\exp(\tfrac{i\pi z(x_v^2+y_v^2)}{\lambda z_1 z_2})\mathrm{Rect}(\tfrac{x_v}{x_v^{max}},\tfrac{y_v}{y_v^{max}}) \\ &\quad \times \mathrm{FFT}^{sgn(z_1)}[u_1(m_1,n_1)\exp(\tfrac{i\pi(x_1^2+y_1^2)}{\lambda z_1})]]. \end{aligned} \tag{2.86}$$

The rectangular function is introduced for the band-limiting chirp function $\exp(\frac{i\pi z(x_v^2+y_v^2)}{\lambda z_1 z_2}) = \exp(2\pi i\phi(x_v,y_v))$ because the result of the first SSF can be regarded as the frequency domain. Aliasing will occur in the absence of the rectangular function. We determine the band-limiting area as follows:

$$1/p_{x_v} \geq 2|f_x^{max}| = 2|\frac{\partial\phi(x_v,y_v)}{\partial x_v}| = |\frac{2zx_v^{max}}{\lambda z_1 z_2}| \tag{2.87}$$

$$1/p_{y_v} \geq 2|f_y^{max}| = 2|\frac{\partial\phi(x_v,y_v)}{\partial y_v}| = |\frac{2zy_v^{max}}{\lambda z_1 z_2}|. \tag{2.88}$$

3 Hologram calculation

Computer-generated holograms (**CGHs**) are used in many applications. For example, CGHs are used in three-dimensional (3D) displays, projection, optical tweezers, beam generation, and multi-spot generation. CGHs are generated by numerical diffraction calculation that projects a two-dimensional (2D) or 3D object on a hologram plane. Calculating holograms from complicated 3D objects in real time is a difficult task even with current computers. In this chapter, we introduce algorithms to calculate holograms at high speed.

3.1 OVERVIEW OF HOLOGRAPHIC DISPLAY

The generation of 3D displays using the CGH technique is called **holographic display**.[i] Holographic 3D displays are attractive because the reconstructed object light is almost the same as the original object light. Figure 3.1 shows the outline of a holographic 3D display system. The computer calculates holograms from 3D object data, and then the hologram is displayed on a **spatial light modulator** (**SLM**). Finally, three-dimensional images can be reproduced in the space by irradiating the SLM with a reconstruction light.

The holographic display can faithfully reconstruct the light of a three-dimensional object so that it can satisfy all human depth perception (binocular parallax, motion parallax, congestion, adjustment, etc.) and is accepted as an ideal three-dimensional display. This is difficult with other three-dimensional display technologies. In principle, if we develop the holographic display system of Figure 3.1, we can realize an ideal one. However, the following problems are hampering practical application.

- The range (**viewing area**) where the observer can see the three-dimensional image is narrow (**viewing angle** is several degrees and difficult for binocular vision).
- The size of the reconstructed three-dimensional image (**field of view**) is small (about several centimeters).
- Hologram calculation time is enormous.

3.1.1 VIEWING ANGLE AND VIEWING AREA

The problem of the viewing angle and field of view is due to the performance of the SLM. Figure 3.2 shows how a three-dimensional object looks when a hologram is displayed on the SLM.

When a reconstruction light is irradiated on this SLM, the diffracted light is generated from the SLM as shown in Figure 3.2(a). When we look through

[i]Also known as, **electroholography**.

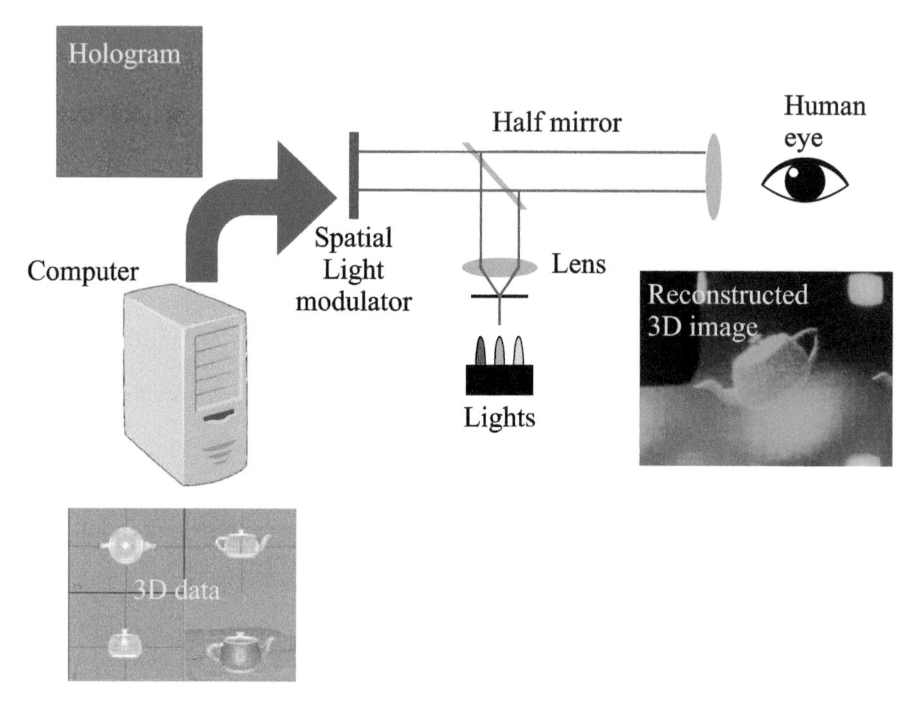

FIGURE 3.1 Holographic display system.

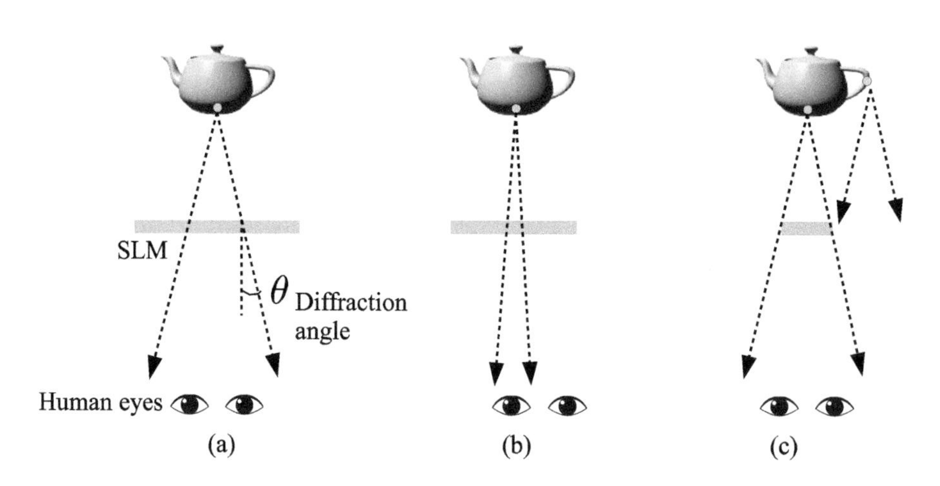

FIGURE 3.2 Viewing angle and field of view.

this diffracted light, we can see a certain part of the object behind the SLM. As shown in Figure 3.2(a), when the diffraction angle θ of the diffracted light is wide, we can binocularly view that three-dimensional object. On the other hand, when θ is small, as shown in Figure 3.2(b), the light can only reach the monocular view, so that we cannot binocularize three-dimensional objects. In addition, simply shifting the eyes, the light does not reach the eyes. Therefore, it becomes a three-dimensional image with a narrow viewing area. This angle is called the **viewing angle**.

Even if the viewing angle is wide as shown in Figure 3.2(c), if the SLM is small, only a part of the object can be observed. The observable size of the object is called the **field of view**. For a holographic display with a wide field of view and a wide viewing angle, an SLM with a large diffraction angle and a large area is required. The diffraction angle of the SLM is determined by the wavelength λ, the pixel pitch p of the SLM, and the incidence angle θ_r of the reconstruction light. The diffraction angle is determined by

$$\sin\theta - \sin\theta_r = \frac{\lambda}{2p}. \tag{3.1}$$

Liquid crystal displays (LCDs) are often used as a hologram display device. At present, the pixel pitch of the LCD used for holographic display is around 4 μm. With such an LCD, the diffraction angle expands to about 3.6° (7.2° on both sides), and binocular vision is possible. As described above, hologram display devices are required to be extremely high definition. Practically, it is said that a pixel pitch of 1 μm or less is required. As the pixel pitch is miniaturized, the viewing angle can be made large. However, the area of the hologram decreases. Therefore, since the hologram is considered to be a window for observing the three-dimensional space behind it, only a part of the three-dimensional image can be observed even if the viewing angle is large. In this way, the ideal display device for holograms is a device with a small pixel pitch and large area.

Table 3.1 shows a summarization of SLMs for holographic display. SLMs have been conducted using an **acoustic optical modulator (AOM)**, a **digital micro-mirror device (DMD)** applying a micro-machining technology, and a **photorefractive material**. The feature of each SLM is as follows.

LCD: This is most frequently used as a hologram display device.

DMD: Display device using micromachining technology. The refresh rate is fast.

AOM: A high-definition hologram can be displayed, but because it is a one-dimensional device, only a one-dimensional hologram can be displayed. The reconstructed image is the only parallax in the horizontal (or vertical) direction.

Photorefractive material: A high-resolution, large-area hologram can be displayed. The refresh rate is slow.

TABLE 3.1
Spatial light modulators for holographic display.

	LCD	DMD	AOM	Photorefractive material
Pixel pitch	4 μm	10 μm	analog	analog
Gradation	256	2	analog	analog
Frame rate	60 Hz	10 KHz	N.A.	slow

Unfortunately, SLMs that can reconstruct 3D images with practical viewing angle and image size have not been developed.

3.1.2 HOLOGRAPHIC DISPLAY SYSTEMS THAT ENLARGE THE VIEWING ANGLE AND FIELD OF VIEW

Here, we introduce holographic display systems to solve the problems of the viewing angle and the field of view as mentioned earlier. In these systems, the viewing angle is enlarged by making smaller the pixel pitch of SLMs, and the field of view is enlarged by enlarging the display area of SLMs.

HOLOGRAPHIC DISPLAY USING THE ACTIVE TILING TECHNIQUE

A schematic diagram of a holographic display using **active tiling** [33, 34] is shown in Figure 3.3(a). Active tiling consists of a ferroelectric LCD ($1,024 \times 1,024$ pixels, 2.5 KHz refresh rate) panel, a copy optical system, shutters, and an **optically addressed SLM** (**OASLM**). The OASLM has high resolution and a large display area, whereas the refresh rate is around 30 Hz.

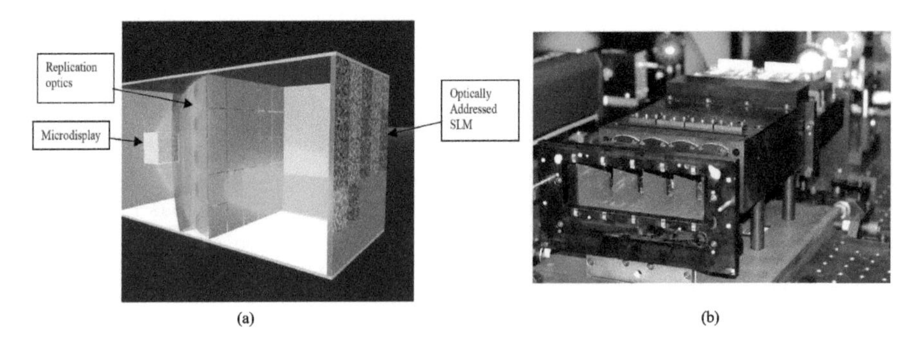

FIGURE 3.3 Schematic diagram of a holographic display using active tiling. Reprinted from Ref. [33] with permission, SPIE.

This method writes small holograms on the OASLM in a time-division manner and finally displays a large hologram on the OASLM. Small holograms are time-sequentially displayed on the ferroelectric LCD panel, and the

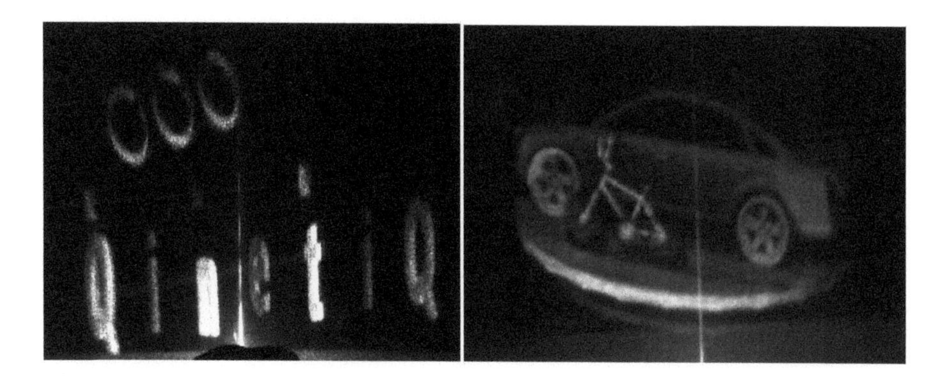

FIGURE 3.4 Reconstructed images from active tiling.
Reprinted from Ref. [33] with permission, SPIE.

hologram is optically copied to a 5 × 5 hologram by the copying optical system. By selecting one of the 5 × 5 holograms with the shutter, a part of the hologram is written to the OASLM at the rear stage. By repeating the same operation 25 times, a large hologram is displayed on the OASLM.

Figure 3.3(b) is a module constructed based on the optical system in Figure 3.3(a). This module can display holograms with 26 million pixels. Furthermore, a hologram with 100 million pixels or more can be reconstructed by a unit combining the four modules. The 3D reconstruction image using this unit is shown in Figure 3.4.

HOLOGRAPHIC DISPLAY WITH PHOTOREFRACTIVE POLYMER

Photorefractive materials have a characteristic that the refractive index changes according to the intensity distribution of incident light. The recorded refractive index can be erased by continuing to irradiate light. Therefore, it can be used as a rewritable recording material. Its resolution has several thousand lines/mm, exceeding electronic devices such as LCDs.

Photorefractive materials include **photorefractive crystals** made of inorganic materials and **photorefractive polymers** made of organic materials. Photorefractive crystals are used in holographic memory, but they are not suitable for display applications because they require a large area, whereas, photorefractive polymers will be able to make large area display devices at low cost.

Reference [35] presents a photorefractive polymer-based hologram display device with a size of 4 inches × 4 inches (about 10 cm × 10 cm). Figure 3.5 shows how to record a hologram to this device and reconstruct the three-dimensional image from the device, and how to erase it [35, 36].

Photorefractive polymers do not have a pixel structure like LCDs and cannot electrically display holograms. Therefore, first of all, holograms are dis-

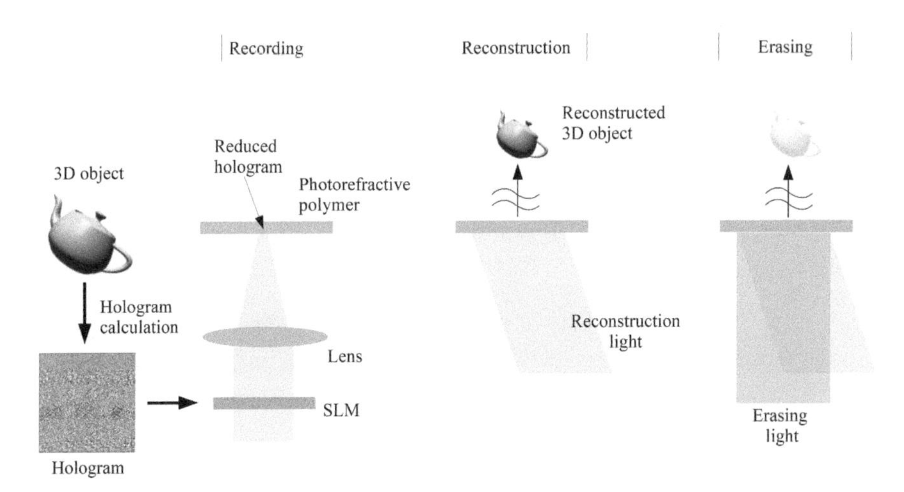

FIGURE 3.5 Recording, reconstruction, and erasing of photorefractive polymers.

played on an LCD, and they are optically transferred onto the photorefractive polymer through the lens. Since the pixel pitch of the LCD is usually coarser than the resolution of the photorefractive polymer, the hologram is reduced by the lens and then recorded on the photorefractive polymer. This makes it possible to record a hologram with a fine pixel pitch, resulting in a reconstructed image with a wide viewing angle.

Since holograms that can be recorded at one time are part of the photorefractive polymers, we will record large holograms while shifting the photorefractive polymer. In order to reconstruct a three-dimensional image from the hologram recorded on the photorefractive polymer, as in ordinary holograms, a reconstruction light is irradiated. When recording a new hologram, it is necessary to erase the recorded hologram once. We can erase this by continuing to irradiate uniform light.

The viewing area is 45 degrees and it is quite wide. The hologram recorded on the photorefractive polymer can continue to observe the image even after three hours. Regarding erasing holograms, we can hardly recognize reconstructed images by continuing to irradiate uniform light for 2 minutes. There is a disadvantage that it takes time to record and erase holograms. However, since photorefractive polymers can observe large three-dimensional images in a wide viewing area, future improvements are highly expected.

3.1.3 PROBLEMS FOR COMPUTATIONAL COST

Another problem of holographic displays is that it is difficult to generate holograms in real time because of the complexity of hologram computation. Figure 3.1 shows the calculation steps of a hologram. The calculation step of the hologram can be divided into the generation/acquisition of three-dimensional

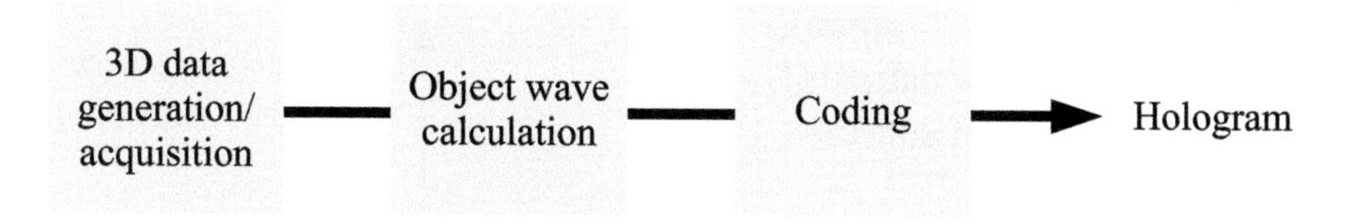

FIGURE 3.6 Calculation steps of a hologram.

object data, calculation of object light from three-dimensional objects, and coding of the object light.

3D data generation / acquisition
Three-dimensional object data can be generated as computer graphics using a three-dimensional graphics library (e.g., OpenGL) and rendering software (e.g., Blender), or obtained by capturing an actual object using a three-dimensional camera (e.g., Kinect).

Object wave calculation
The distribution of light emanating from the 3D object data is calculated on a hologram plane.

Coding
The calculated object light has a complex amplitude. Because the complex amplitude cannot normally be displayed directly on SLMs, we convert the complex amplitude to display on the SLMs. This conversion is called **coding**.

We want to realize a practical holographic display by calculating the hologram from 3D objects in real time. However, the practical use of CGH-based 3D displays is hampered by the long computational time required for diffraction calculation. For example, a CGH calculation with the CGH resolution of $1{,}024 \times 1{,}024$ using a ray-tracing algorithm from about 52,000 object points took 300 s on Intel Core2 Quad Q6600 processor.

In the calculation step of Figure 3.1, object light calculation has the greatest complexity. The following methods are mainly used for the calculation of object light.

- Point cloud method
- Polygon method
- RGB-D method
- Multi-view image method

These methods have advantages and disadvantages. These methods can be used singly or in combination. In this section, we introduce the acceleration of these calculation methods.

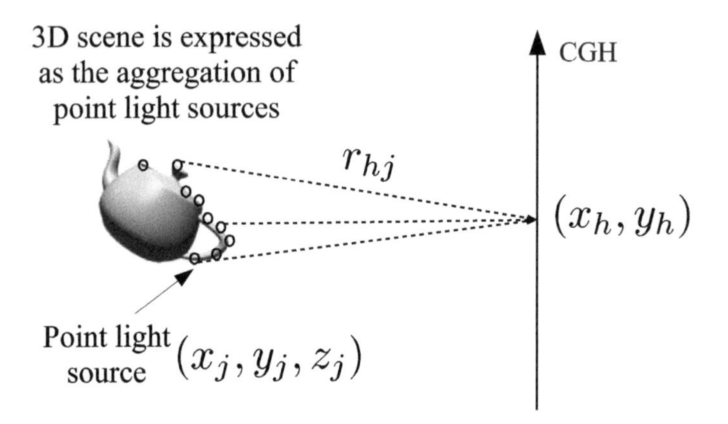

FIGURE 3.7 Point cloud-based CGH calculation.

3.2 POINT CLOUD METHOD

Figure 3.7 shows a **point cloud**-based CGH calculation. The point cloud approach expresses a 3D scene as aggregated **point light sources** (**PLSs**). It can be thought that a PLS emits a spherical wave. Therefore, the complex amplitude $u(x_h, y_h)$ on the hologram plane can be calculated by adding up the spherical waves emitted by the PLSs. The calculation is quite simple, and is given by

$$u(x_h, y_h) = \sum_{j=1}^{M} \frac{A_j}{r_{hj}} \exp(ikr_{hj}) \tag{3.2}$$

where $i = \sqrt{-1}$, $u(x_h, y_h)$ is the complex amplitude on the hologram, M is the number of PLSs, A_j is the light intensity of the jth PLS, and r_{hj} indicates the distance between the jth PLS and (x_h, y_h) in the hologram. r_{hj} is calculated as

$$r_{hj} = \sqrt{(x_h - x_j)^2 + (y_h - y_j)^2 + z_j^2}. \tag{3.3}$$

By converting (coding) the complex amplitude u into an amplitude hologram, this hologram can be displayed in an amplitude SLM. The amplitude hologram $I(x_h, y_h)$ is calculated by taking the real part of the complex amplitude as follows:

$$I(x_h, y_h) = \mathrm{Re}\{u(x_h, y_h)\} = \sum_{j=1}^{M} \frac{A_j}{r_{hj}} \cos(kr_{hj}). \tag{3.4}$$

By converting (coding) the complex amplitude u into a **kinoform**, this kinoform can be displayed on a phase-modulated SLM. The kinoform $theta(x_h, y_h)$ is calculated by taking the phase of the complex amplitude as follows:

$$\theta(x_h, y_h) = \tan^{-1} \frac{\mathrm{Im}\{u(x_h, y_h)\}}{\mathrm{Re}\{u(x_h, y_h)\}}. \tag{3.5}$$

For the hologram size of $N_h \times N_h$, the computational complexity is $O(N_h^2 M)$; therefore, fast calculation algorithms are required.

3.2.1 LOOK-UP TABLE ALGORITHMS

The point cloud algorithm of Eq. (3.2) is most commonly implemented by using a **look-up table (LUT)**. The time-consuming computations in the equation are those for the square root and exp functions; therefore, a simple LUT method [37] precalculates $\exp(kr_{hj})$ of Eq. (3.2) and stores it in the LUT. Subsequently, when accumulating light sources, we simply read the precalculated data from the LUT. In this method, it is necessary to prepare the LUT as

$$L(x_h, y_h, x_j, y_j, z_j) = \exp(ikr_{hj}), \tag{3.6}$$

for each object point, so that a lot of memory is required. We can calculate Eq. (3.2) using the LUT by

$$u(x_h, y_h) = \sum_{j=1}^{M} A_j L(x_h, y_h, x_j, y_j, z_j). \tag{3.7}$$

Although the LUT successfully accelerates the CGH calculation, it requires a large memory allocation because the LUT needs to be prepared for every (x_j, y_j, z_j).

N-LUT

Recently, the large memory requirement of Eq. (3.6) has been reduced by the **novel LUT (N-LUT)** method [38–40]. In the N-LUT, the depth of the 3D object is discretized and the zone plate is calculated by

$$L(x_h, y_h, z_j) = \exp(ikr_{hj}). \tag{3.8}$$

For the lateral direction (x_j, y_j) of the object point, we only shift these zone plates. As a result, it is sufficient to store the zone plates only in the depth direction, so the amount of memory can be reduced. Subsequently, the N-LUT is calculated by

$$u(x_h, y_h) = \sum_{j=1}^{M} A_j L(x_h - x_j, y_h - y_j, z_j). \tag{3.9}$$

S-LUT

Since the N-LUT needs to hold the zone plate in two dimensions, it still uses a large amount of memory. The **split look-up table (S-LUT)** can reduce the LUT to one dimension by separating the zone plate into horizontal and vertical components [41].

Using **Fresnel approximation**, Eq. (3.2) can be written as:

$$u(x_h, y_h) = \sum_{j=1}^{M} A_j \exp(i\frac{2\pi}{\lambda}(z_j + \frac{(x_h - x_j)^2 + (y_h - y_j)^2}{2z_j})). \tag{3.10}$$

Adding $\frac{2\pi z_j}{\lambda}$ to the phase of Eq. (3.10), Eq. (3.10) can be written as

$$u(x_h, y_h) = \sum_{j=1}^{M} A_j \exp(i\frac{2\pi}{\lambda}(z_j + \frac{(x_h - x_j)^2}{2z_j} + z_j + \frac{(y_h - y_j)^2}{2z_j})). \tag{3.11}$$

This addition is an integer multiple of 2π, so it does not affect the calculation result. Defining L_H and L_V as

$$L_H(x_h - x_j, z_j) = \exp(i\frac{2\pi}{\lambda}(z_j + \frac{(x_h - x_j)^2}{2z_j})), \tag{3.12}$$

$$L_V(y_h - y_j, z_j) = \exp(i\frac{2\pi}{\lambda}(z_j + \frac{(y_h - y_j)^2}{2z_j})), \tag{3.13}$$

Eq. (3.11) is rewritten as

$$u(x_h, y_h) = \sum_{j=1}^{M} A_j L_H(x_h - x_j, z_j) L_V(y_h - y_j, z_j). \tag{3.14}$$

L_H and L_V represent LUTs storing one-dimensional zone plates.

C-LUT

Since the S-LUT only needs to prepare a one-dimensional zone plate, the memory amount can be reduced as compared with the N-LUT. However, it is necessary to prepare a lot of one-dimensional zone plates when the depth of a three-dimensional object is deep. The **compressed LUT (C-LUT)** [42] is a method that can reduce the number of zone plates in the depth direction by using **Fraunhofer approximation**.

The hologram can be calculated by using

$$u(x_h, y_h) = \sum_{j=1}^{M} A_j \exp(i\frac{2\pi}{\lambda}(\frac{(x_h - x_j)^2 + (y_h - y_j)^2}{2z_j})). \tag{3.15}$$

As described in Section 2.4, under the condition of Fraunhofer approximation, we can use the following approximation:

$$\exp(i\frac{2\pi}{\lambda}(x_j^2 + y_j^2)) \approx 1. \tag{3.16}$$

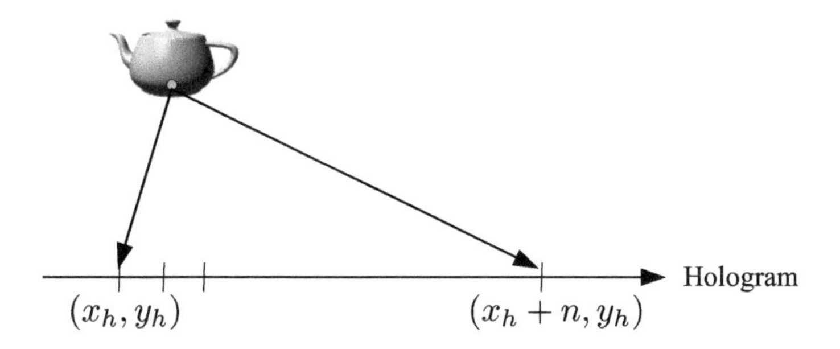

FIGURE 3.8 Point cloud-based CGH calculation method.

Applying this approximation to Eq. (3.15), we write Eq. (3.15) as

$$\begin{aligned} u(x_h, y_h) &= \sum_{j=1}^{M} A_j \exp(i\frac{2\pi}{\lambda}(\frac{x_h^2 + y_h^2}{2z_j} - (\frac{x_h x_j + y_h y_j}{z_j})), \\ &= \sum_{j=1}^{M} A_j \exp(i\frac{2\pi}{\lambda}(\frac{x_h^2 + y_h^2}{2(d+\Delta_j)} - (\frac{x_h x_j + y_h y_j}{d+\Delta_j})), \end{aligned} \tag{3.17}$$

where we defined d as the reference point in one of the object points, Δ_j as the depth deviation between object points, and used $z_j = d + \Delta_j$.

Assuming $\Delta_j \ll d$, Eq. (3.18) is written as

$$\begin{aligned} u(x_h, y_h) &= \sum_{j=1}^{M} A_j \exp(\frac{2\pi i}{\lambda}(\frac{x_h^2 + y_h^2}{2(d+\Delta_j)} - (\frac{x_h x_j + y_h y_j}{d})), \\ &= \sum_{j=1}^{M} A_j \exp(\frac{2\pi i}{\lambda}(\frac{x_h^2 + y_h^2}{2(d+\Delta_j)}) \exp(-\frac{2\pi i x_h x_j}{\lambda d}) \exp(-\frac{2\pi i y_h y_j}{\lambda d}), \\ &= \sum_{j=1}^{M} A_j L_H(x_h, x_j) L_V(y_h, y_j) L_L(\Delta_j), \end{aligned} \tag{3.18}$$

where

$$L_H(x_h, x_j) = \exp(-i\frac{2\pi x_h x_j}{\lambda d}), \tag{3.19}$$

$$L_V(y_h, y_j) = \exp(-i\frac{2\pi y_h y_j}{\lambda d}), \tag{3.20}$$

$$L_L(\Delta_j) = \exp(i\frac{2\pi}{\lambda}(\frac{x_h^2 + y_h^2}{2(d+\Delta_j)}). \tag{3.21}$$

3.2.2 RECURRENCE ALGORITHM

Recurrence-relation-based hologram calculations have been proposed [43–46]. This section briefly describes the recurrence algorithm based on [46], which is especially well suited to hardware implementation.

The hologram calculation under Fresnel approximation is expressed as

$$u(x_h, y_h) = \sum_{j=1}^{M} A_j \exp(i\frac{2\pi}{\lambda}(z_j + \frac{(x_h - x_j)^2 + (y_h - y_j)^2}{2z_j})). \quad (3.22)$$

The recurrence algorithm can compute the phase of Eq. (3.22) by recurrence relation.

Figure 3.8 shows the arrangement of a 3D object and a hologram. The phase θ_H at $(x_h,\ y_h)$ on the hologram plane propagated from $(x_j,\ y_j)$ in a 3D object is expressed by

$$\theta_H(x_{hj},\ y_{hj},\ z_j) = 2\pi(\frac{z_j}{\lambda} + \frac{p^2}{2\lambda z_j}(x_{hj}^2 + y_{hj}^2)) = 2\pi(\theta_Z + \theta_{XY}), \quad (3.23)$$

where x_{hj} and y_{hj} denote $(x_h - x_j)$ and $(y_h - y_j)$, respectively. p is the sampling interval on the hologram plane. The coordinates (x_h, y_h) and (x_j, y_j) are normalized by p.

θ_{XY} and θ_Z are defined as

$$\theta_{XY}(x_{hj},\ y_{hj},\ z_j) = \frac{p^2}{2\lambda z_j}(x_{hj}^2 + y_{hj}^2), \quad (3.24)$$

$$\theta_Z(z_j) = \frac{z_j}{\lambda}. \quad (3.25)$$

The phase θ_H is computed by θ_{XY} and θ_Z. We consider the phase $\theta_{XY}(x_{hj} + d,\ y_{hj},\ z_j)$ at the position $(x_h + d,\ y_h)$ on the hologram. The phase $\theta_{XY}(x_{hj} + d,\ y_{hj},\ z_j)$ is expressed by

$$\theta_{XY}(x_{hj} + d,\ y_{hj},\ z_j) = \frac{p^2}{2\lambda z_j}((x_{hj} + d)^2 + y_{hj}^2)$$

$$= \frac{p^2}{2\lambda z_j}(x_{hj}^2 + y_{hj}^2) + \frac{p^2}{2\lambda z_j}(2dx_{hj} + d^2) = \theta_{XY}(x_{hj},\ y_{hj},\ z_j) + \Gamma_d. \quad (3.26)$$

Here, Γ_d is defined as

$$\Gamma_d(x_{hj},\ z_j) = \frac{p^2}{2\lambda z_j}(2dx_{hj} + d^2). \quad (3.27)$$

We substitute $1, 2, 3, \cdots, n$ into d on Γ_d.

For $d = 1$, Γ_1 is expressed by

$$\Gamma_1 = \frac{p^2}{2\lambda z_j}(2x_{hj} + 1). \quad (3.28)$$

For $d = 2$, Γ_2 is expressed by

$$\Gamma_2 = \frac{p^2}{2\lambda z_j}(4x_{hj} + 4) = \frac{p^2}{2\lambda z_j}(2x_{hj} + 1) + \frac{p^2}{2\lambda z_j}(2x_{hj} + 1) + \frac{p^2}{2\lambda z_j} \times 2$$

$$= \Gamma_1 + \Gamma_1 + \Delta. \tag{3.29}$$

Here, Δ is defined as

$$\Delta(z_j) = \frac{p^2}{2\lambda z_j} \times 2. \tag{3.30}$$

For $d = 3$, Γ_3 is expressed by

$$\Gamma_3 = \frac{p^2}{2\lambda z_j}(6x_{hj} + 9) = \frac{p^2}{2\lambda z_j}(4x_{hj} + 4) + \frac{p^2}{2\lambda z_j}(2x_{hj} + 1) + \frac{p^2}{2\lambda z_j} \times 4$$

$$= \Gamma_2 + \Gamma_1 + 2\Delta. \tag{3.31}$$

For $d = n$, we can commonly express it as

$$\Gamma_n = \Gamma_{n-1} + \Gamma_1 + (n-1)\Delta. \tag{3.32}$$

Here, we put the second and third terms as δ_{n-1}. Therefore, we rewrite Eq. (3.32) as follows.

$$\Gamma_n = \Gamma_{n-1} + \delta_{n-1}. \tag{3.33}$$

Here, δ_{n-1} is expressed by

$$\delta_{n-1} = \Gamma_1 + (n-1)\Delta. \tag{3.34}$$

δ_n at the next coordinate is expressed by

$$\delta_n = \Gamma_1 + n\Delta. \tag{3.35}$$

When we subtract Eq. (3.34) from Eq. (3.35), δ_n is expressed by

$$\delta_n = \delta_{n-1} + \Delta. \tag{3.36}$$

Eventually, we can compute the phase Γ_n at the next coordinate using two recurrence formulas (Eq. (3.33) and Eq. (3.36)). Therefore, we can compute the phase θ_H at the position $(x_h + n,\ y_h)$ on a hologram by adding Γ_n and θ_Z to θ_{XY}.

In the case that we compute a hologram to use this method with a fixed-point operation, we can ignore the overflow bit caused by addition in the process of all phase computations, which are from Eq. (3.23) to Eq. (3.36). Because the overflow bit is regarded as one period of the cosine function, the dynamic ranges of θ_Z, θ_{XY}, Γ_1, Δ, Γ_n and δ_n are always constant, respectively.

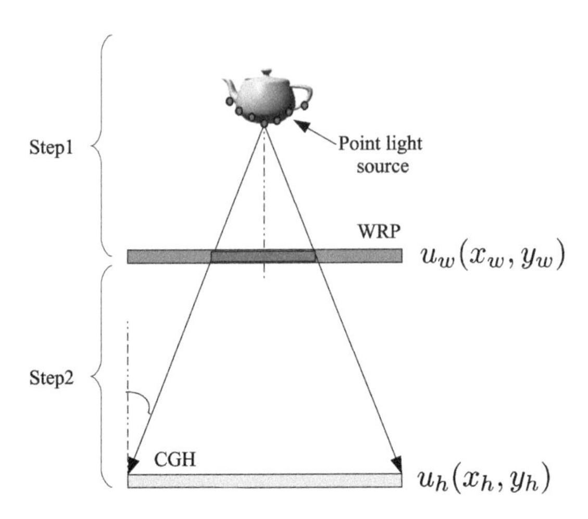

FIGURE 3.9 Wavefront recording plane method.

3.2.3 IMAGE HOLOGRAM ALGORITHM AND WAVEFRONT RECORDING PLANE METHOD

In an **image hologram** as described in Section 1.2.2, a 3D object is placed in the vicinity of the hologram. When the 3D object is placed close to the hologram, light emitted from the object point spreads only in a small area of the hologram. Therefore, it is not necessary to calculate the light distribution of the entire surface of the hologram, which reduces the amount of calculation.

An image hologram algorithm enables real-time 3D color reconstruction using a CPU. This approach calculates a small region of light from a 3D object close to a hologram, rather than the entire region on the hologram [47,48]. On an Intel Core 2 Quad CPU with three CPU threads, this algorithm executed at a rate of 19 fps for a 1,400 $\times$ 1,050 pixel CGH constructed from 10,000 PLSs. The image hologram approach reduces the computational complexity to $O(\bar{W}^2 M)$, where $\bar{W}$ is the average radius of the zone plates on the hologram generated by each object point; unfortunately, this method cannot reconstruct a 3D object in the Fresnel region.

Inspired by the image hologram method, **wavefront recording plane (WRP)** methods have been proposed [49–56]. As shown in Figure 3.9, this method proceeds in two steps: in the first step, a virtual plane (WRP $u_w(x_w, y_w)$) is placed between a 3D object and a hologram and the WRP is calculated by

$$u_w(x_w, y_w) = \sum_{j=1}^{M} \frac{A_j}{r_{wj}} \exp(ikr_{wj}) \tag{3.37}$$

where r_{wj} is the distance between the object point and the WRP; in the second step, diffraction calculations from the WRP to the hologram are performed as

$$u_h(x_h, y_h) = \text{Prop}_z[u_w(x_w, y_w)]. \tag{3.38}$$

If the WRP is placed near a 3D object, the computational cost of the first step can be reduced; therefore, the total computational complexity is dramatically lower than that in conventional hologram calculations. The WRP method reduces the computational complexity of a hologram to $O(\bar{W}^2 M + N_h^2 \log N_h)$.

3.2.4 WAVELET TRANSFORM-BASED ALGORITHM

To further accelerate the point cloud-based hologram calculation, **WAvelet ShrinkAge-Based superpositIon (WASABI)** has been proposed [57–59]. Wavelet shrinkage [60, 61] eliminates small wavelet coefficient values of the zone plate u_{z_j} thereby resulting in an approximated zone plate calculated from a few representative wavelet coefficients. The zone plate is calculated by

$$u_{z_j}(x_h, y_h) = \exp(i\frac{2\pi}{\lambda}\sqrt{x_h^2 + y_h^2 + z_j^2}). \tag{3.39}$$

The use of these representative wavelet coefficients accelerates the superposition of zone plates in the wavelet domain.

The superposition calculation illustrated in Eq. (3.2) is time-consuming. The objective of using WASABI is to superpose the zone plate as quickly as possible. The WASABI calculations are performed as follows:

1. Precomputation: we precompute the zone plate $u_{z_j}(x_h, y_h)$ and then transform the space-domain zone plate into a wavelet-domain zone plate. The number of wavelet coefficients is reduced by wavelet shrinkage.
2. Superposition: the reduced wavelet coefficients are superposed in the wavelet domain.
3. Inverse transformation: the conversion back from the wavelet domain to the space domain is performed.

PRECOMPUTATION

In the precomputation step, a zone plate, $u_{z_j}(x_h, y_h)$, is calculated from an object point with $(0, 0, z_j)$ and is then transformed from the space-domain zone plate into the wavelet domain using a **fast wavelet transform (FWT)** [60]. The computational complexity of the forward and inverse FWTs is $O(N_h^2)$.

According to wavelet theory, a zone plate, $u_{z_j}(x_h, y_h)$, is decomposed using scaling coefficients $s_{m,n}^{(\ell)}$ as follows:

$$u_{z_j}(x_h, y_h) = \sum_{m=-\infty}^{\infty}\sum_{n=-\infty}^{\infty} s_{m,n}^{(0)}\phi(x_h - m)\phi(y_h - n), \tag{3.40}$$

where $m, n \in \mathbb{Z}$, ℓ denotes the level of the wavelet decomposition, and ϕ is the scaling function. According to the wavelet theory, we do not need to know the mathematical form of a scaling function. Instead, we need to know the two-scale sequences in Eq. (3.42). Mallat suggested that the zero level of the scaling coefficients can be considered as the original signal [62] such that $u_{z_j}(x_h, y_h)$ can be expressed as

$$u_{z_j}(x_h, y_h) \approx s^{(0)}_{m,n}. \tag{3.41}$$

FWT decomposes $u_{z_j}(x_h, y_h)$ into scaling coefficients $s^{(\ell)}_{m,n}$ and wavelet coefficients $w^{(\ell)}_{LH,m,n}$, $w^{(\ell)}_{HL,m,n}$, $w^{(\ell)}_{HH,m,n}$ by

$$\begin{aligned}
s^{(\ell+1)}_{m,n} &= \sum_{k_2}\sum_{k_1} p_{k_1-2m} p_{k_2-2n} s^{(\ell)}_{m,n}, \\
w^{(\ell+1)}_{LH,m,n} &= \sum_{k_2}\sum_{k_1} p_{k_1-2m} q_{k_2-2n} s^{(\ell)}_{m,n}, \\
w^{(\ell+1)}_{HL,m,n} &= \sum_{k_2}\sum_{k_1} q_{k_1-2m} p_{k_2-2n} s^{(\ell)}_{m,n}, \\
w^{(\ell+1)}_{HH,m,n} &= \sum_{k_2}\sum_{k_1} q_{k_1-2m} q_{k_2-2n} s^{(\ell)}_{m,n},
\end{aligned} \tag{3.42}$$

where p and q are two-scale sequences. We recursively repeat Eq. (3.42) until the desirable level $\ell = L$ is achieved. Several two-scale sequences have been proposed previously. In Ref. [57], we use Daubechies 4 [60] for the two-scale sequences and chose $L = 2$ because we empirically found that these settings result in good-quality reconstructed images. The two-scale sequences are as follows:

$$\begin{aligned}
p_k &= \{0.23037781330885523, 0.7148465705525415, \\
&\quad 0.6308807679295904, -0.02798376941698385, \\
&\quad -0.18703481171888114, 0.030841381835986965, \\
&\quad 0.032883011666982945, -0.010597401784997278\} \\
q_k &= \{-0.010597401784997278, -0.032883011666982945, \\
&\quad 0.030841381835986965, 0.18703481171888114, \\
&\quad -0.02798376941698385, -0.6308807679295904, \\
&\quad 0.7148465705525415, -0.23037781330885523\}
\end{aligned}$$

After the FWT, the scaling and wavelet coefficients are generally distributed as shown in Figure 3.10. Figure 3.10(c) shows the zone plate in the wavelet domain (Figure 3.10(b)) whose calculation conditions are $z_j = 100$ mm, a pixel pitch of 10 μm and a wavelength of 633 nm. The number of raw scaling and wavelet coefficients is the same as the number of pixels in the hologram plane, N_h^2.

If the next step (the superposition step) uses the raw coefficients, we cannot expect an acceleration of the hologram calculation; therefore, we use wavelet

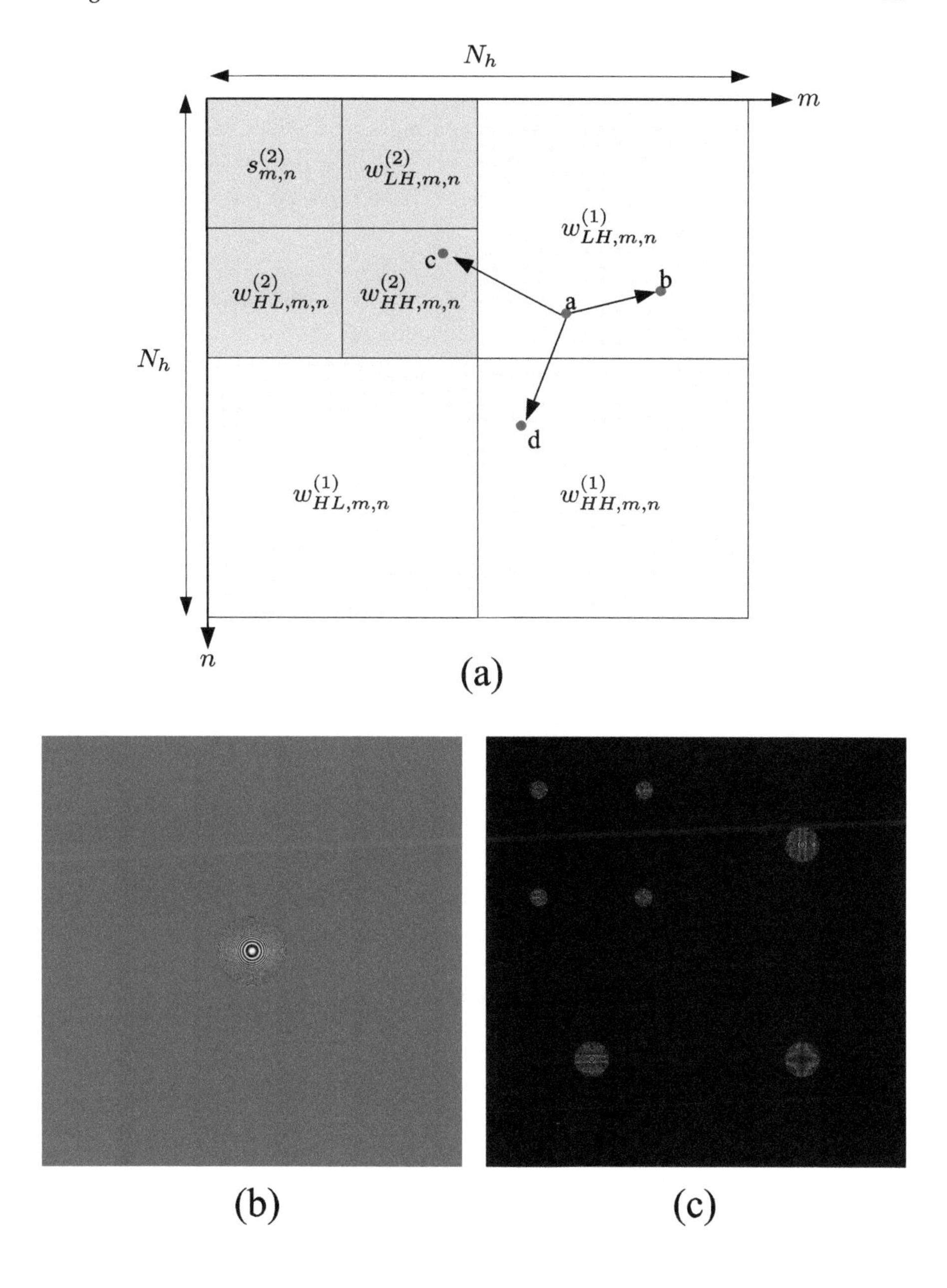

FIGURE 3.10 Distribution of the scaling and wavelet coefficients: (a) distribution of the scaling and wavelet coefficients, (b) a zone plate, and (c) the zone plate in the wavelet domain.

shrinkage, which eliminates small wavelet coefficient values to express the approximated zone plate using a few representative wavelet coefficients.

After sorting the amplitude of the scaling and wavelet coefficients $|s_{m,n}^{(\ell)}|$ and $|w_{m,n}^{(\ell)}|$ where $|a| = \sqrt{\mathrm{Re}\{a\} \times \mathrm{Re}\{a\} + \mathrm{Im}\{a\} \times \mathrm{Im}\{a\}}$, in the descending order of amplitude we select N_r coefficients from the larger amplitudes. In this paper, N_r is defined as

$$N_r = 2\pi {W_j}^2 \times r, \tag{3.43}$$

where W_j is the radius of a zone plate u_{z_j}, and r is the selection rate among the larger amplitudes. The selected coefficients are stored in vector $\mathbf{v}_z$, which is described as follows:

$$\mathbf{v}_z = [c_{z,0}, \alpha_{z,0}, \cdots, c_{z,N_r-1}, \alpha_{z,N_r-1}], \tag{3.44}$$

where

$$c_{z,k} \in \{s_{m,n}^{(\ell)}, w_{LH,m,n}^{(\ell)}, w_{HL,m,n}^{(\ell)}, w_{HH,m,n}^{(\ell)}\}, \tag{3.45}$$

takes any one of $s_{m,n}^{(\ell)}$, $w_{LH,m,n}^{(\ell)}$,$w_{HL,m,n}^{(\ell)}$ and $w_{HH,m,n}^{(\ell)}$ for a PSF u_{z_j}; $\alpha_{z,k}$ is a shift weight of $c_{z,k}$ used in the superposition step and is expressed as

$$\alpha_{z,k} = 2^{-\ell}. \tag{3.46}$$

As shown in Figure 3.10(a), the level ℓ has different values depending on (m, n).

The number of vectors is N_z, where N_z is the number of z-direction slices for the 3D objects and the vectors are stored in a LUT for the numerical calculation.

SUPERPOSITION

In the superposition step, we superpose the reduced coefficients related to all the object points in the wavelet domain. For example, if we calculate an object point with (x_1, y_1, z_1), we first read the reduced coefficients of u_{z_1} from the vectors $\mathbf{v}_z$. The reduced coefficients $c_{z,k}$ at a position (m, n) are then shifted to $2^{-\ell}x_1 = \alpha_{z,k}x_1$ and $2^{-\ell}y_1 = \alpha_{z,k}y_1$ in the horizontal and vertical directions, and the shifted coefficients are added to the wavelet space $\psi(m, n)$. The mathematical expression of the superposition is as follows:

$$\psi(m, n) = \sum_{j=0}^{N} a_j \sum_{k=0}^{N_r-1} c_{z_j,k}\delta(m - \alpha_{z_j,k}x_j, n - \alpha_{z_j,k}y_j), \tag{3.47}$$

where $\delta(m, n)$ is the Dirac delta function. Note that the shifted $c_{z_j,k}$ should be in the same region as the original $c_{z_j,k}$. For example, in Figure 3.10(a), if the point "a" of $c_{z_j,k}$ is shifted to point "b," the shifted $c_{z_j,k}$ can be superposed using Eq. (3.47) because the region of these points is the same. Conversely, if the point "a" of $c_{z_j,k}$ is shifted to point "c" or "d," the shifted $c_{z_j,k}$ should not be superposed because the region of these points differs from that of point "a." The computational complexity of Eq. (3.47) is $O(r\bar{W}^2N)$, which is clearly less than that obtained with Eq. (3.2).

INVERSE TRANSFORMATION

In the inverse transformation step to obtain the complex amplitude in the hologram plane, the superposed coefficients $\psi(m,n)$ are transformed from the wavelet domain to the space domain using inverse FWT. We extract new wavelet coefficients $s^{(\ell)}_{m,n}$, $w^{(\ell)}_{LH,m,n}$,$w^{(\ell)}_{HL,m,n}$ and $w^{(\ell)}_{HH,m,n}$ from $\psi(m,n)$; then the inverse FWT process reconstructs the lower level of scaling coefficients from the higher level of scaling and wavelet coefficients as

$$s^{(\ell)}_{m,n} = \sum_{k_2}\sum_{k_1} p_{m-2k_2}p_{n-2k_1}s^{(\ell+1)}_{m,n} + p_{m-2k_2}q_{n-2k_1}w^{(\ell+1)}_{LH,m,n}$$

$$+q_{m-2k_2}p_{n-2k_1}w^{(\ell+1)}_{HL,m,n} + q_{m-2k_2}q_{n-2k_1}w^{(\ell+1)}_{HH,m,n}. \tag{3.48}$$

Eventually, from Eq. (3.41), we can consider the zero level of scaling coefficients $s^{(0)}_{m,n}$ obtained using Eq. (3.48) as

$$u(x_h, y_h) = \sum_{j}^{N} a_j u_{z_j}(x_h - x_j, y_h - y_j, z_j) \approx s^{(0)}_{m,n}. \tag{3.49}$$

The computational complexity of the inverse FWT is $O(N_h^2)$. Finally, we convert the complex amplitude to the amplitude hologram, or to kinoform.

After precomputation, we do not need to perform the precomputation step again during the hologram calculation. We only repeat the superposition and inverse transformation steps with respect to new 3D objects; therefore, the computational complexity of the proposed method is expressed as

$$O(r\bar{W}^2N) + O(N_h^2). \tag{3.50}$$

For example, if we set $r = 1\%$, we can reduce the computational complexity of the superposition up to two orders of magnitude.

Figures 3.11(a)-3.11(c) show an original PSF and the approximated PSFs using WASABI with $z_j = 0.1$ m and $r = 10\%$, and 5%. Figures 3.11(d)-3.11(f) show the interference patterns of two object points using Eq. (3.2) and WASABI with $r = 10\%$, and 5%. The result demonstrates the generation of the interference for the two points via the wavelet domain.

PERFORMANCE

We compare the computational performance of WASABI and the quality of images reconstructed using it with those of the **N-LUT** [39]. The calculation conditions involved a wavelength of 633 nm, a hologram sampling interval of 10 μm, and a hologram resolution of 2,048 *times* 2,048 pixels. The depth range of the 3D images was 5 mm, equally divided into 10 steps ($N_z = 10$).

Figure 3.12 shows the reconstructed images, image quality measured by peak signal-to-noise ratios (PSNRs) and calculation times of a merry-go-round

Zone plate

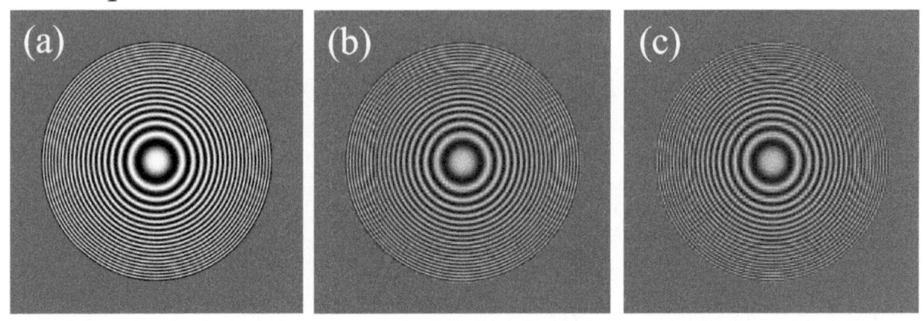

Interference fringes

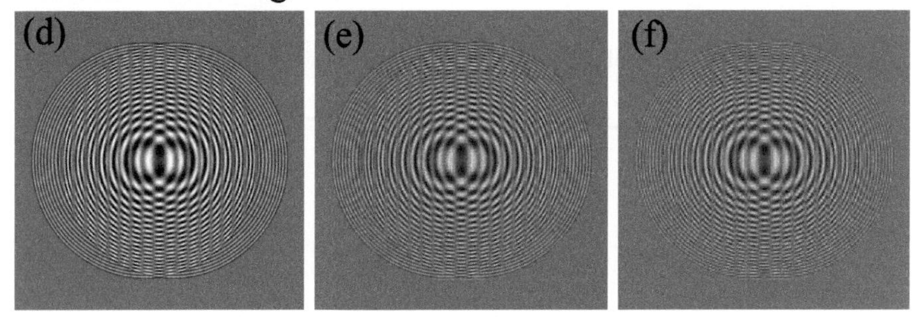

FIGURE 3.11 Point spread functions: (a) the original PSF and (b) approximated PSFs with $r = 10\%$, and c) $r = 5\%$. Interference fringes constructed from two object points using (d) the original PSF and (e) approximated PSFs with $r = 10\%$ and (f) $r = 5\%$.

composed of approximately 100,000 points.[ii] The reconstructed images in the top row of Figure 3.12 were obtained using the conventional method (N-LUT) at two different distances of 0.5 m and 0.05 m. The reconstructed images in the middle row were obtained using WASABI with $r = 5\%$. The PSNRs of the reconstructed images to the conventional method are 30.6 dB at a distance of 0.5 m and 30.2 dB at a distance of 0.05 m, respectively. The reconstructed images in the bottom row were obtained via WASABI with $r = 1\%$, and the PSNRs of the reconstructed images are 29.6 dB at a distance of 0.5 m and 29.4 dB at a distance of 0.05 m, respectively. The PSNRs obtained using WASABI with $r = 1\%$ are lower than those obtained using WASABI with $r = 5\%$; however, the calculation time is further accelerated. For the distance of 0.5 m, as shown in Figure 3.12, the calculation time for the conventional method, WASABI with $r = 5\%$, and WASABI with $r = 1\%$ are approximately 281, 44, and 7 s, respectively. WASABI with $r = 1\%$ provides an approximately 41-fold improvement in speed over the conventional method. In addition, at a distance of 0.05 m, WASABI can calculate the holograms at approximately 5 fps using the CPU.

3.3 POLYGON APPROACH

Figure 3.13 illustrates a polygon-based hologram calculation. In the polygon approach, 3D scenes are expressed as the aggregation of planar polygons. A hologram is calculated by accumulating diffracted fields of each polygon tilted to the hologram. The accumulation is calculated by

$$u(x_h, y_h) = \sum_{j=1}^{P} u_j(x_h, y_h) \tag{3.51}$$

where P is the number of polygons and $u_j(x_h, y_h)$ is the light distribution on the hologram from j-th polygon. The efficient calculation of tilted diffraction, $u_j(x_h, y_h)$, is crucial in the polygon approach because FFT-based diffraction calculations[iii] as described in Chapter 2 can only be used to compute light propagation between parallel planes. Although tilted diffraction can be calculated directly using Eq. (2.1) and Eq. (2.20) without using FFT, the computational complexity is $O(N^4)$ where the hologram size is $N \times N$ pixels, resulting in a difficult real-time calculation.

To overcome this difficulty, tilted diffraction using FFTs has been proposed [16, 63–67]. These methods, except for Ref. [67], obtain a diffracted field from a tilted polygon by calculating the angular spectrum of the source

[ii] We used a computer with Intel Core i7 6700 as a CPU, a memory of 64 GB, a Microsoft Windows 8.1 operating system, and a Microsoft Visual Studio C++ 2013 compiler. All of the calculations were parallelized across eight CPU threads.

[iii] For example, the angular spectrum method and Fresnel diffraction.

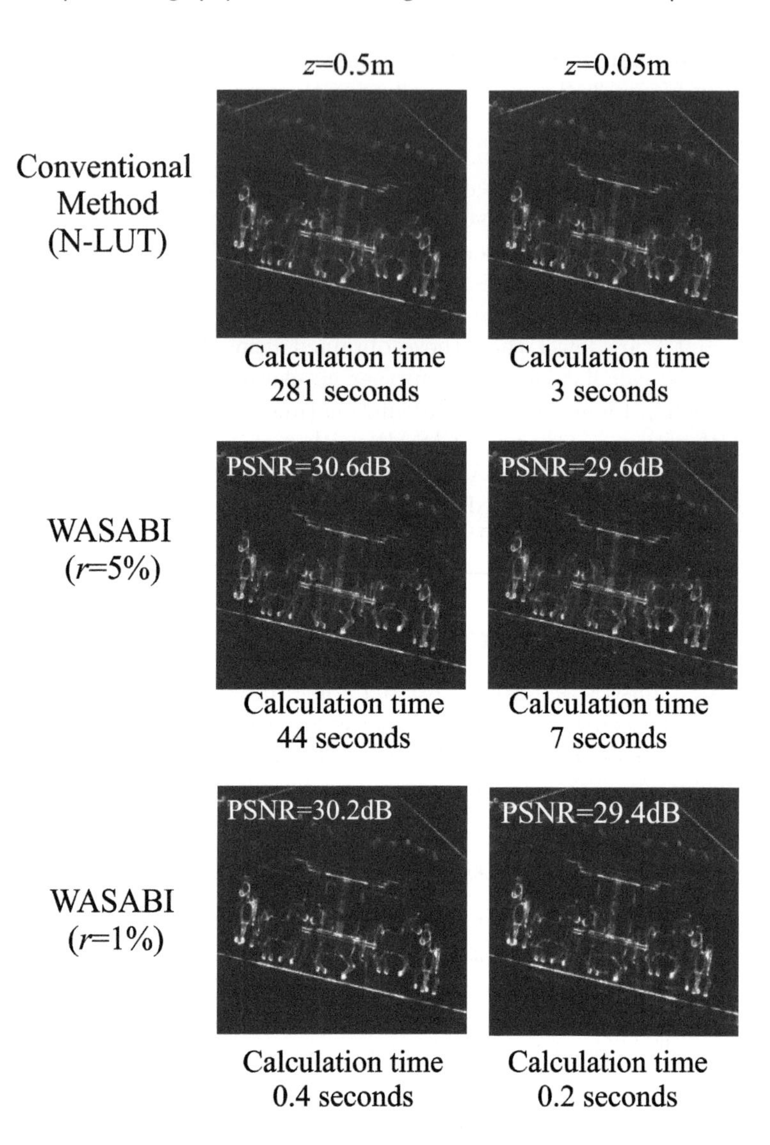

FIGURE 3.12 Reconstructed images, image quality and calculation times at two different distances of 0.5 m and 0.05 m: the merry-go-round composed of 95,949 points.

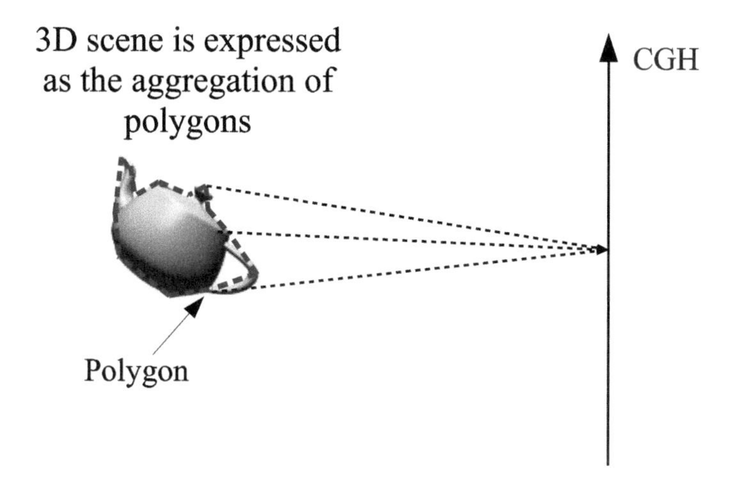

FIGURE 3.13 Polygon-based hologram calculation.

or destination plane, followed by the rotation of the angular spectrum in the frequency domain. Finally, the tilted diffracted field in a spatial domain is obtained using the inverse FFT.

This section explains tilted diffraction using FFTs [16, 63–66].

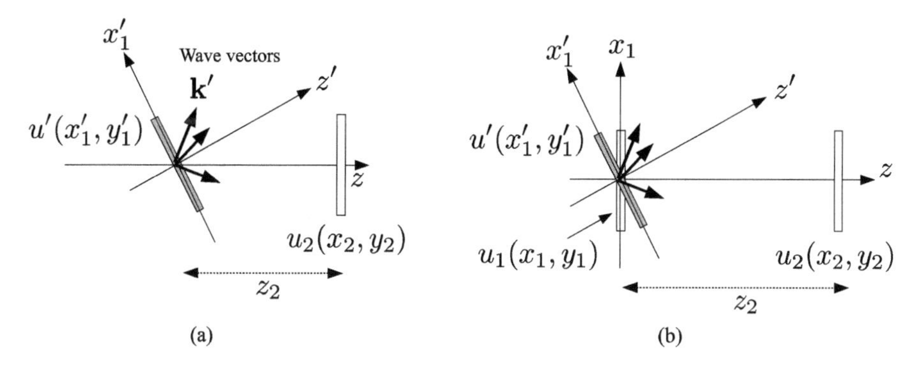

FIGURE 3.14 Tilted diffraction.

As shown in Figure 3.14(a), let us consider how to calculate the propagation from $u'_1(x'_1, y'_1)$, which is perpendicular to the z' axis tilted by a certain angle from the z axis, to $u_2(x_2, y_2)$. This is calculated by the following two steps:

- First step: As shown in Figure 3.14(b), we calculate the light distribution when observing $u'_1(x'_1, y'_1)$ on $u_1(x_1, y_1)$.
- Second step: We calculate the diffraction from $u_1(x_1, y_1)$ to $u_2(x_2, y_2)$. Since $u_1(x_1, y_1)$ and $u_2(x_2, y_2)$ are parallel, the diffraction calculation introduced in Chapter 2 can be used.

Since this tilted diffraction is based on the angular spectrum method, we consider the meaning of the angular spectrum again. The angular spectrum of $u'_1(x'_1, y'_1)$ is calculated by using the Fourier transform as

$$U'_1(f'_x, f'_y) = \mathcal{F}\left[u'_1(x'_1, y'_1)\right], \tag{3.52}$$

where (f'_x, f'_y) is the spatial frequency. $u'_1(x'_1, y'_1)$ is expressed by using the angular spectrum as

$$\begin{aligned} u'_1(x'_1, y'_1) &= \mathcal{F}^{-1}\left[U'_1(f'_x, f'_y)\right] \\ &= \int\int U'_1(f'_x, f'_y)\exp(2\pi i(f'_x x'_1 + f'_y y'_1))df'_x df'_y. \end{aligned} \tag{3.53}$$

On the other hand, a plane wave traveling along the z' axis can be written as

$$a\exp(i\mathbf{k}' \cdot \mathbf{r}') = a\exp(i(k'_x x'_1 + k'_y y'_1 + k'_z z'_1)) \tag{3.54}$$

where a is the amplitude, $\mathbf{k}' = (k'_x, k'_y, k'_z)$ is the wave vector and $\mathbf{r}' = (x'_1, y'_1, z'_1)$ is the position vector.[iv]

Assuming that $u'_1(x'_1, y'_1)$ is placed at $z'_1 = 0$, adding $f'_z z'_1$ to Eq. (3.53) does not matter. Therefore, we can rewrite Eq. (3.53) as

$$\begin{aligned} u'_1(x'_1, y'_1) &= \mathcal{F}^{-1}\left[U'_1(f'_x, f'_y)\right] \\ &= \int\int U'_1(f'_x, f'_y)\exp(2\pi i(f'_x x'_1 + f'_y y'_1 + f'_z z'_1))df'_x df'_y. \end{aligned} \tag{3.55}$$

f'_z represents the spatial frequency in the propagation direction. Comparing Eq. (3.55) and Eq. (3.54), we see that the following relationship holds between the wave vector and the spatial frequency (f'_x, f'_y, f'_z):

$$\mathbf{k}' = (k'_x, k'_y, k'_z) = 2\pi(f'_x, f'_y, f'_z). \tag{3.56}$$

Therefore, $u'_1(x'_1, y'_1)$ is constructed by summing plane waves propagating in various wave vectors with amplitude of $U'_1(f'_x, f'_y)$. In addition, because $|\mathbf{k}'| = \sqrt{k'^2_x + k'^2_y + k'^2_z} = k$ ($k = 2\pi/\lambda$ is the wave number.) is not independent, it is a function of k'_x and k'_y. Therefore, k'_z is expressed as

$$k'_z(k'_x, k'_y) = \sqrt{k'^2 - k'^2_x - k'^2_y}. \tag{3.57}$$

[iv]See Section 1.1.

From Eq. (3.56), f'_z is not independent as well, but is a function of f'_x and f'_y:

$$f'_z(f'_x, f'_y) = \sqrt{\frac{1}{\lambda^2} - f'^2_x - f'^2_y}. \tag{3.58}$$

If the $u'_1(x'_1, y'_1)$ is calculated by the angular spectrum method as it is, it becomes the diffraction calculation along the z' axis. Using a rotation matrix $\mathbf{R}$, there is the following relation between the wave vector $\mathbf{k}'$ propagating to z' axis and $\mathbf{k}$ propagating to the z axis[v]:

$$\mathbf{k}' = \mathbf{R}\mathbf{k}. \tag{3.59}$$

This equation is rewritten as

$$\begin{pmatrix} k'_x \\ k'_y \\ k'_z \end{pmatrix} = \begin{pmatrix} a_{11} & a_{12} & a_{13} \\ a_{21} & a_{22} & a_{23} \\ a_{31} & a_{32} & a_{33} \end{pmatrix} \begin{pmatrix} k_x \\ k_y \\ k_z \end{pmatrix}. \tag{3.60}$$

As shown in Figure 3.15, the rotation matrix $\mathbf{R}$ is $\mathbf{R_x}$ when rotating around the x axis, $\mathbf{R_y}$ when rotating around the y axis, and $\mathbf{R_z}$ when rotating around the z axis.

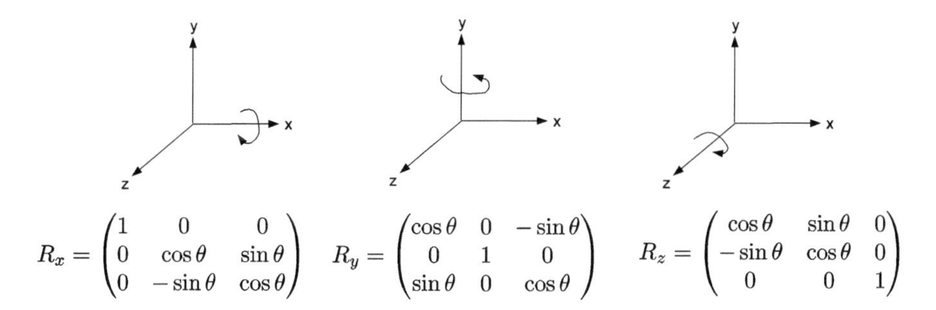

FIGURE 3.15 Rotation matrix.

With the relationship of Eq. (3.56), Eq. (3.60) can be rewritten as

$$\begin{pmatrix} f'_x \\ f'_y \\ f'_z \end{pmatrix} = \begin{pmatrix} a_{11} & a_{12} & a_{13} \\ a_{21} & a_{22} & a_{23} \\ a_{31} & a_{32} & a_{33} \end{pmatrix} \begin{pmatrix} f_x \\ f_y \\ f_z \end{pmatrix}. \tag{3.61}$$

Therefore, we obtain

$$\begin{aligned} f'_x &= \alpha(f_x, f_y) = a_{11}f_x + a_{12}f_y + a_{13}f_z(f_x, f_y) \\ f'_y &= \beta(f_x, f_y) = a_{21}f_x + a_{22}f_y + a_{23}f_z(f_x, f_y). \end{aligned} \tag{3.62}$$

[v]The origin of $u'_1(x'_1, y'_1)$ and $u_1(x_1, y1)$ are common.

f_z is not independent but is a function of f_x and f_y as follows:

$$f_z(f_x, f_y) = \sqrt{\frac{1}{\lambda^2} - f_x^2 - f_y^2}. \tag{3.63}$$

By substituting the coordinates (f_x, f_y) we want into Eq. (3.62), the corresponding coordinates (f'_x, f'_y) can be obtained. Finally, the angle spectrum $U_1(f_x, f_y)$ is calculated by

$$U_1(f_x, f_y) = U'_1(\alpha(f_x, f_y), \beta(f_x, f_y)). \tag{3.64}$$

When numerically calculating Eq. (3.64), it is necessary to pay attention to the following points. The $U_1(f_x, f_y)$ that we want is sampled at even intervals (Figure 3.16(b)).

On the other hand, $U'_1(f'_x, f'_y)$ can be obtained simply by calculating the FFT of $u'_1(x'_1, y'_1)$, but this result is also sampled at even intervals (Figure 3.16(a)). We can find $U_1(f_x, f_y)$ from $U'_1(f'_x, f'_y)$ by substituting (f_x, f_y) into Eq. (3.62).

However, since Eq. (3.62) is nonlinear, the calculated coordinates may be the position of the star in Figure 3.16(a), not equidistant sampling points. In such a case, it is necessary to calculate the angular spectrum obtained by interpolation processing from the surrounding angular spectrum $U'_1(f'_x, f'_y)$.

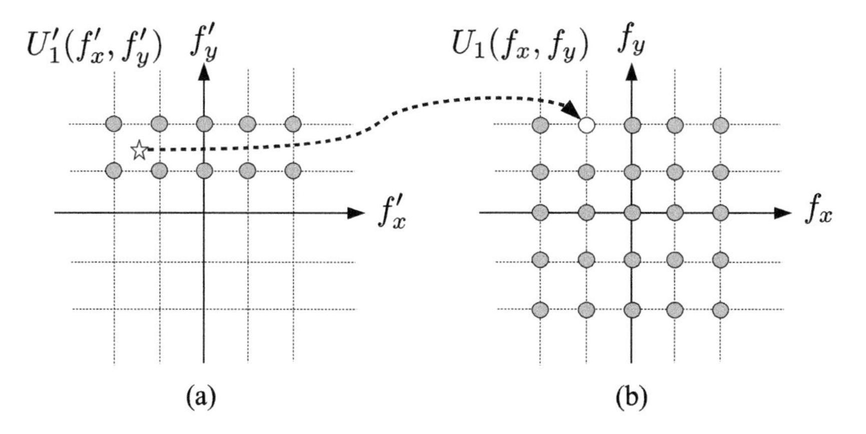

FIGURE 3.16 Interpolation processing for angular spectrum. The star denotes irregular sampling points and the circles denote regular sampling points.

The angular spectrum of $u_1(x_1, y_1)$ parallel to $u_2(x_2, y_2)$ is obtained from Eq. (3.64). Finally, we can obtain the light distribution of $u_2(x_2, y_2)$ by using the angular spectrum method:

$$u_2(x_2, y_2) = \mathcal{F}^{-1}\left[U_1(f_x, f_y)|J(f_x, f_y)| \exp(i2\pi z_2\sqrt{\frac{1}{\lambda^2} - f_x^2 - f_y^2})\right], \tag{3.65}$$

where $|J(f_x, f_y)|$ is Jacobian.[vi] $|J(f_x, f_y)|$ is defined as

$$\begin{aligned}
|J(f_x, f_y)| &= \begin{vmatrix} \frac{\partial \alpha}{\partial f_x} & \frac{\partial \alpha}{\partial f_y} \\ \frac{\partial \beta}{\partial f_x} & \frac{\partial \beta}{\partial f_y} \end{vmatrix} = \left| \frac{\partial \alpha}{\partial f_x}\frac{\partial \beta}{\partial f_y} - \frac{\partial \alpha}{\partial f_y}\frac{\partial \beta}{\partial f_x} \right| \\
&= (a_{11}a_{22} - a_{12}a_{21}) + (a_{12}a_{23} - a_{13}a_{22})\frac{f_x}{f_z(f_x, f_y)} + \\
&\quad (a_{13}a_{21} - a_{11}a_{23})\frac{f_y}{f_z(f_x, f_y)}. \qquad (3.66)
\end{aligned}$$

Since $U_1'(f_x', f_y')$ and $U_1(f_x, f_y)$ only change viewpoints in different coordinate systems, these energies should be equal. Therefore, the following equation holds:

$$\int\int |U_1'(f_x', f_y')|^2 df_x' f_y' = \int\int |U_1'(\alpha(f_x, f_y), \beta(f_x, f_y))|^2 |J(f_x, f_y)| df_x df_y. \qquad (3.67)$$

Figure 3.17 shows the calculation results of $u_1(x_1, y_1)$ when rotating $u_1'(x_1', y_1')$ at 30° and 70° around the y axis, respectively.

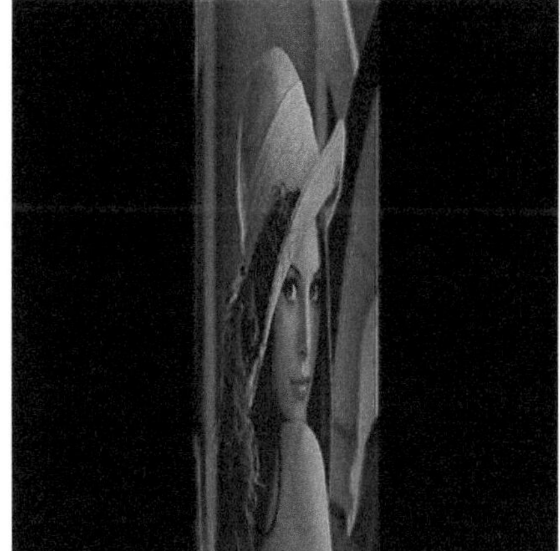

FIGURE 3.17 Rotating result using the tilted diffractionThe left image is the result for the rotation angle of 30°The right image is the result for the rotation angle of 70°

Figure 3.18 shows the 3D reconstructed image calculated from a 3D object with 718 polygons [1]. The hologram size is $65,536 \times 65,536$ pixels. This hologram is produced on a quartz substrate with a chromium film using a laser lithography system and etching the chromium film after laser drawing.

[vi] Since the integral area expands and reduces before and after the variable transformation of Eq. (3.62), Jacobian is a term for canceling the expansion / reduction.

Since the pixel pitch is 1 μm and the size of the hologram is about 6.5 cm $\times$ 6.5 cm, when we observe the hologram from various directions, we can see that 3D reconstructed images with different amounts of parallax can be obtained.

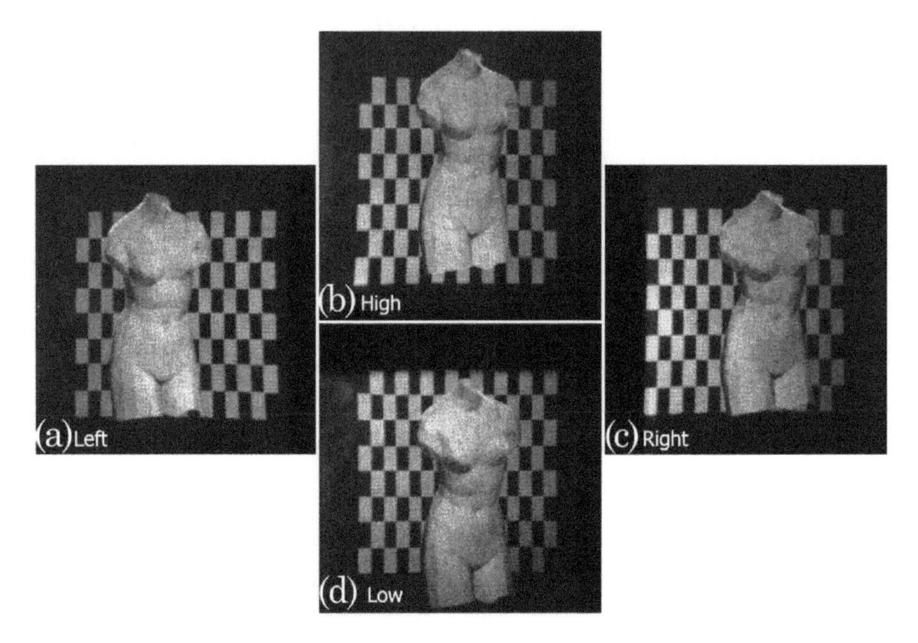

FIGURE 3.18 Reconstructed images from a hologram calculated from polygon model.Reprinted from Ref. [1] with permission, OSA.

3.3.1 MULTI-VIEW IMAGE APPROACH

Figure 3.19 shows a schematic of a multi-view image-based hologram calculation. The **multi-view image approach** calculates a hologram from 2D images captured at different viewing positions on the hologram plane. Unlike point-cloud and polygon approaches, this approach requires neither hidden surface removal nor shading processing because both are automatically performed by the computer graphics technique. Moreover, this approach readily manages natural 3D scenes and 3D computer graphics. Multi-view image approaches can be subcategorized into **holographic stereograms** and **integral photography**-based hologram calculations.

Holographic stereograms [68] are composites of elemental holograms (**hogels**) generated from 2D images by FFT. The computational cost of holographic stereograms is much lower than that of the point-cloud and polygon approaches. On the other hand, although an observer can recognize a 3D image reconstructed from elemental holograms, the quality of deep depth images is problematic. This is because of the fact that the reconstructions are essentially 2D images. To improve this method, researchers have proposed phase-added

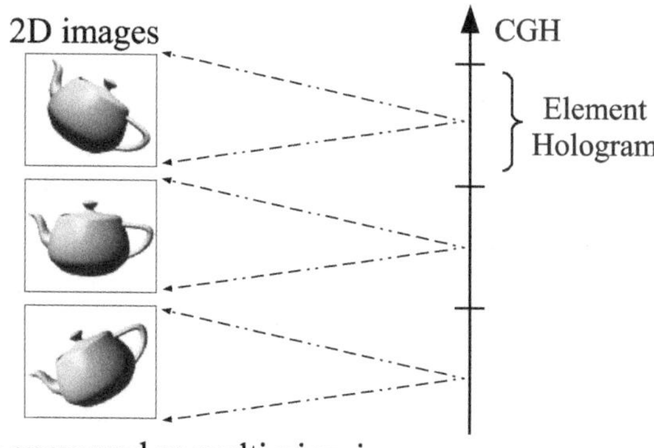

FIGURE 3.19 Multiview image-based hologram calculation.

stereogram [69] and accurate phase-added stereogram approaches [70]. Computational costs of original and modified holographic stereogram are almost identical. Recently, a simple holographic stereogram calculation method using spherical converging light has been proposed [71].

Integral photography (IP), invented by Lippmann in 1908, collects elemental images of natural 3D scenes using a lens array and then reconstructs the 3D scenes through the same lens array. IP-based hologram calculation has been proposed [72]; this method captures 3D scenes through the IP technique and converts elemental images to elemental holograms. Assisted by a graphics processing unit (GPU), Ref. [73] recently generated holograms with 7,680 × 4,320 pixels at 12 frames per second from elemental images captured by the IP technique.

We now explain how to calculate a hologram by the multi-view image approach. First, we generate multi-view images in some way (Figure 3.20). Representative multi-view image generation methods include a method of moving a camera, using a lens array, and using a plurality of cameras (camera array). If we physically use lens arrays and cameras, we can capture real-world multi-view images. On the other hand, if we move a virtual camera in 3D computer graphics, we can generate a multi-view image of the 3D graphics.

Here, we use OpenGL, the well-known 3D graphics library, to generate multi-view images. We divide the camera plane into small areas and capture 3D objects from the center of the small area with the camera. An image taken from a small area is called a sub-image. We move the camera to the next small area and take a sub-image. By repeating this, it is possible to obtain a set of sub-images captured at different positions.

The movement of the camera uses the gluLookAt function in OpenGL.

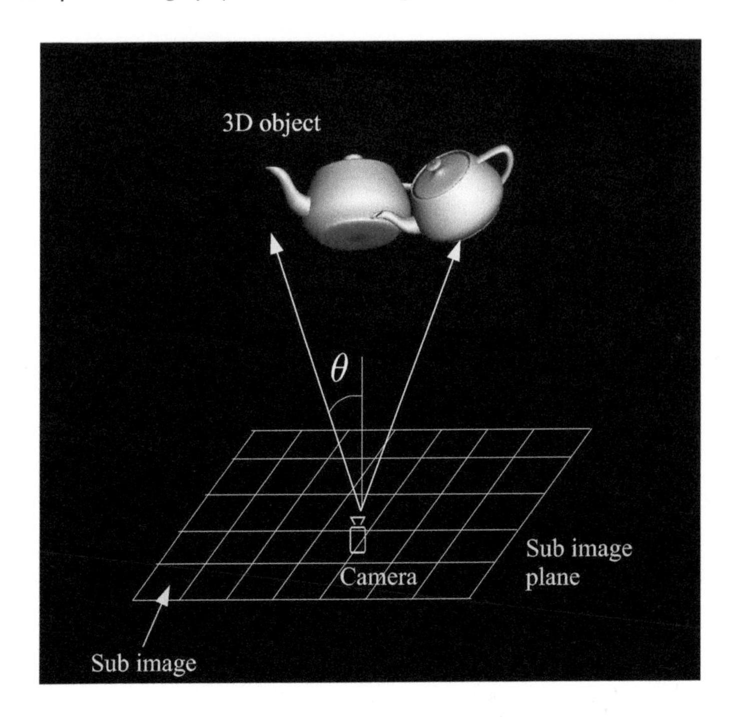

FIGURE 3.20 Generation of sub-images.

The angle of view of the camera, 2θ, depends on the diffraction angle of the hologram. Therefore, when the sampling interval of the hologram is p, it becomes $\theta = \sin^{-1}\frac{\lambda}{2p}$. This angle of view is set in the glutPerspective function, and the sub-image is acquired by perspective projection. Figure 3.21(a) is a set of sub-images taken with different viewpoints. Figure 3.21(c) is the enlarged view. Next, we convert the set of the sub-images (Figure 3.21(a)) into a set of element images (Figure 3.21(b)) [74].[vii] Figure 3.21(d) is the enlarged view.

We can calculate the j-th elemental hologram (hogel) $u_j(m_2, n_2)$ by Fourier transform (FFT) of the j-th elemental image $u_j(m_1, n_1)$. $u_j(m_2, n_2)$ can be calculated by

$$u_j(m_2, n_2) = \text{FFT}\left[u_j(m_1, n_1)\exp(i2\pi n(m_1, n_1))\right], \tag{3.68}$$

where $\exp(i2\pi n(m_1, n_1))$ represents the random phase [viii] where $n(m_1, n_1)$ is a random value ranging from 0 to 1. By computing all hogels $u_j(m_2, n_2)$, we can calculate the whole hologram $u(m_2, n_2)$.

[vii]The IP technique using a lens array can directly capture element images.

[viii]The reason why the random phase is required will be explained in Section 5.3.4.

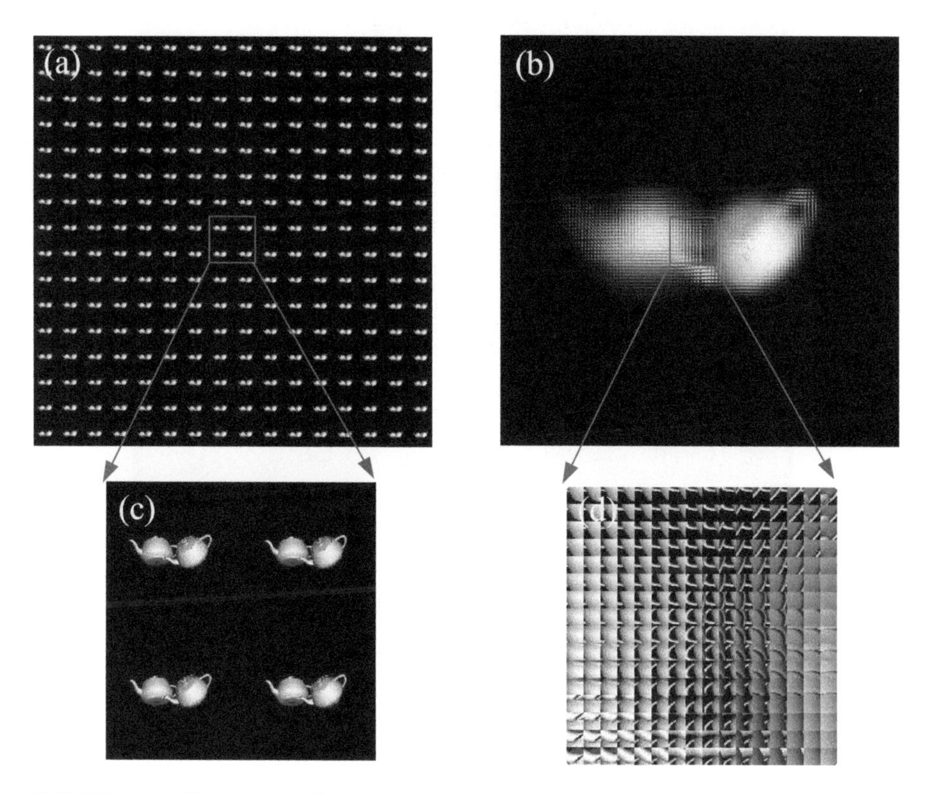

FIGURE 3.21 Conversion of sub-images to elemental images: (a) sub-images and (b) elemental images.

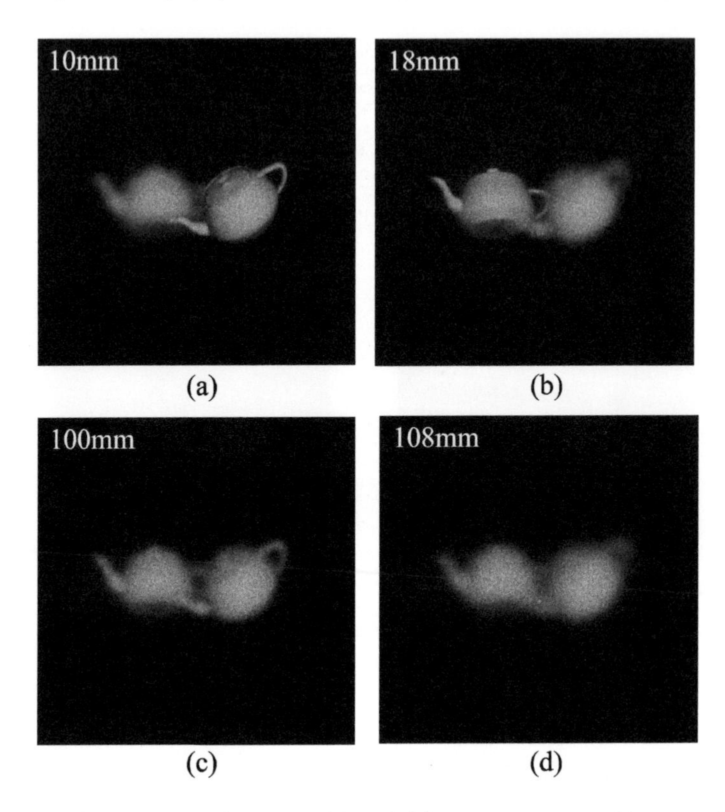

FIGURE 3.22 Reconstructed images: (a) and (b) are reconstructed images of the 3D scene located near from the hologram. (c) and (d) are reconstructed images of the 3D scene located far from the hologram.

Figures 3.22(a) and (b) are reconstructed images when a 3D scene is placed near the hologram. In the 3D scene, two teapots were placed 10 mm and 18 mm from the hologram. When the reconstructed image is near the hologram, the image quality of the reconstructed image is good.

On the other hand, Figures 3.22(c) and (d) are the reconstructed images of the 3D scene located far from the hologram. The two teapots were placed 100 mm and 108 mm from the hologram. The image quality of the reconstructed image is deteriorated. As with integral photography, the reconstructed image from the hologram calculated by the multi-view image approach has a problem that the image quality of the reconstructed image deteriorates when it is far from the hologram.

Deep 3D scenes reconstructed by the multi-view image approach can be improved using the **ray-sampling method** (**RS method**) [75]. When the distance between the three-dimensional object and the hologram is long, a virtual plane (ray sampling plane) is arranged in the vicinity of the 3D object and elemental images are acquired on the virtual plane. Similarly to Eq. (3.68),

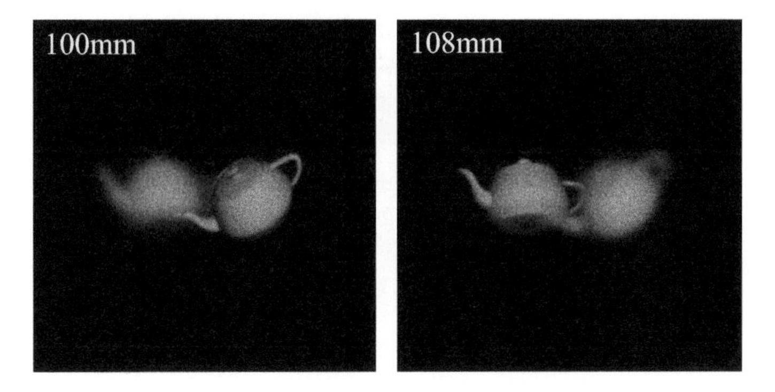

FIGURE 3.23 Reconstructed images obtained by the RS method.

we calculate the FFT of all the elemental images to calculate the light wave on the virtual plane. The result is denoted as $u_r(m_r, n_r)$.

Finally, the hologram $u_2(m_2, n_2)$ is calculated by performing the diffraction calculation from the virtual plane to the hologram plane. This calculation is performed by

$$u_2(m_2, n_2) = \text{Prop}_z\left[u_r(m_r, n_r)\right]. \tag{3.69}$$

Figure 3.23 is the reconstructed image from the hologram calculated by the RS method. Comparing with Figures 3.22(c) and (d), we can see that the image quality of the reconstructed image is improved.

In the above explanation, the method of generating a hologram from a set of element images has been described, but a method of generating a hologram from a set of sub-images (Figure 3.21(a)) has also been proposed. In addition, instead of using perspective projection, a method for generating holograms using orthogonal projection has been proposed [76, 77].

Methods that can reproduce deep 3D scenes by combining the multi-view image approach with other methods have been proposed. For example, Refs. [78, 79] combine a set of sub-images and depth images, and generate a hologram with the point cloud method without using FFT. In addition, Refs. [80, 81] combine the RGB-D image approach described in the next section and the multi-view image approach to create a hologram. Using integral photography, we can convert a live-action 3D scene into a hologram. Other methods using a light field camera have also been proposed [82, 83].

3.3.2 RGB-D IMAGE APPROACH

Figure 3.24 shows a schematic of an RGB-D image-based hologram calculation.[ix] **RGB-D images** are composites of RGB images and depth images. Today, RGB-D images are readily obtained by commercial RGB-D cameras

[ix] This method is also know as a **layer-based method**.

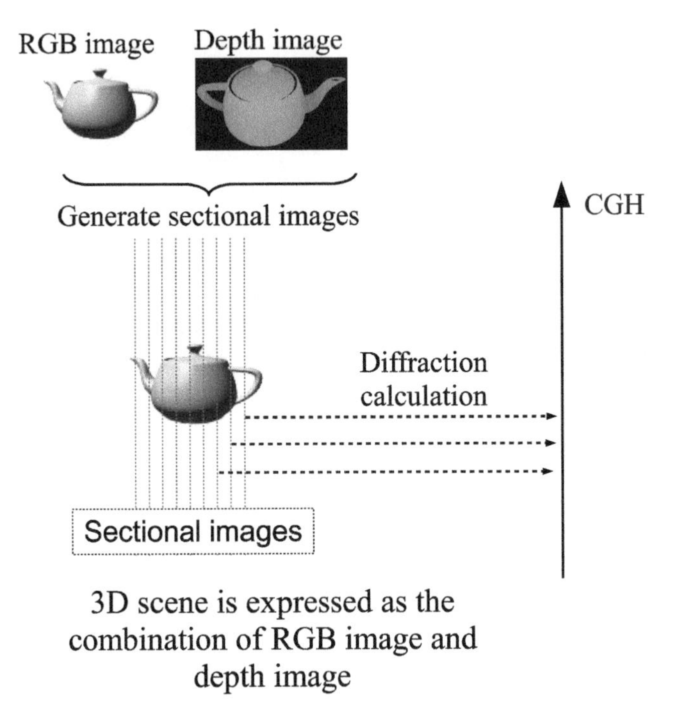

FIGURE 3.24 RGB-D image-based hologram calculation.

such as Microsoft Kinect and also from 3D graphics data through 3D graphics libraries such as OpenGL. A hologram is generated from an RGB-D image as follows [84]:

$$u_1(m_2, n_2) = \sum_{j=0}^{D} \text{Prop}_j\{\text{rgb}(m_1, n_1)\exp(i2\pi n(m_1, n_1))\text{mask}_j(m_1, n_1)\}, \tag{3.70}$$

where j is the depth index, D is the depth range (usually 255), Prop_j denotes the diffraction calculation of the j-th propagation distance, $\text{rgb}(m_1, n_1)$ is the RGB image, and $\exp(i2\pi n(m_1, n_1))$ is the random phase[x] where $n(m_1, n_1)$ is selected from a uniform distribution of pseudorandom numbers ranging from 0.0 to 1.0. The function $\text{mask}_j(m_1, n_1)$ is defined as

$$\text{mask}_j(m_1, n_1) = \begin{cases} 1 & (if \;\; \text{dep}(m_1, n_1) = j), \\ 0 & (otherwise), \end{cases} \tag{3.71}$$

where $\text{dep}(m_1, n_1)$ is the depth image (ranging from 0 to 255). Here, a diffraction calculation is the most time-consuming part. Diffraction methods adopted in the RGB-D approach include **band-limited double-step**

[x]The reason why the random phase is required will be explained in Section 5.3.4.

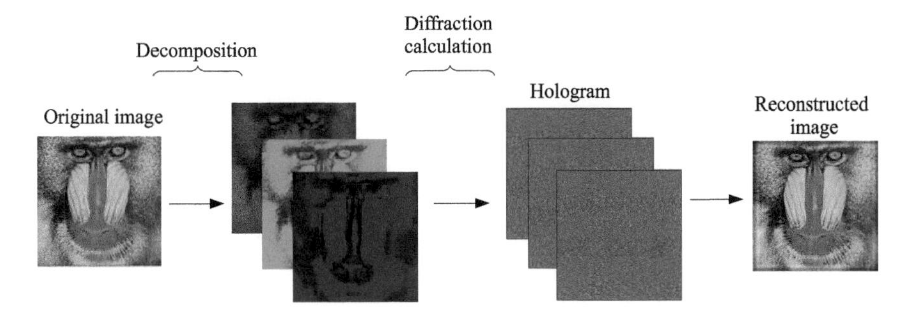

FIGURE 3.25 Color CGH calculation in RGB color space.

Fresnel diffraction [32] and **double-step Fresnel diffraction** [23]. This diffraction calculation is explained in Section 2.8.2. The latter reduces memory requirements and calculation time but introduces aliasing that degrades a diffracted field. Band-limited double-step Fresnel diffraction not only improves the aliasing problem but also executes 2-3 times faster than the methods following the angular spectrum method Eq. (2.11) and convolution-based Fresnel diffraction Eq. (2.24).

3.3.3 ACCELERATION OF COLOR HOLOGRAM CALCULATION

Color hologram calculations are generally calculated in the RGB color space as shown in Figure 3.25, although a faster color CGH algorithm has been developed in [85–87] by converting an RGB image to a YCbCr color space as shown in Figure 3.26, where Y, Cb, and Cr denote the luminance, blue-difference chroma, and red-difference chroma components, respectively. The human eye recognizes tiny differences in the luminance component but is less sensitive to the chroma components.

In the YCbCr color space, the luminance component is normally sampled, whereas the chroma components are downsampled. The downsampling accelerates the calculation of color holograms. The Y, Cb, and Cr components are processed by diffraction calculations, and the diffracted results are then converted to the RGB color space. Theoretically, this method can be up to three times faster than calculations performed in the RGB color space. The method also accelerates the calculation of the color hologram from RGB-D images [86]. Figure 3.27 shows reconstruction images obtained from color holograms calculated by the transformation from the RGB to the YCbCr color space.

3.3.4 OTHER ALGORITHMS

Reference [88] compared the calculation times of a point cloud approach with recurrence relation and FFT-based polygon approaches. They concluded that

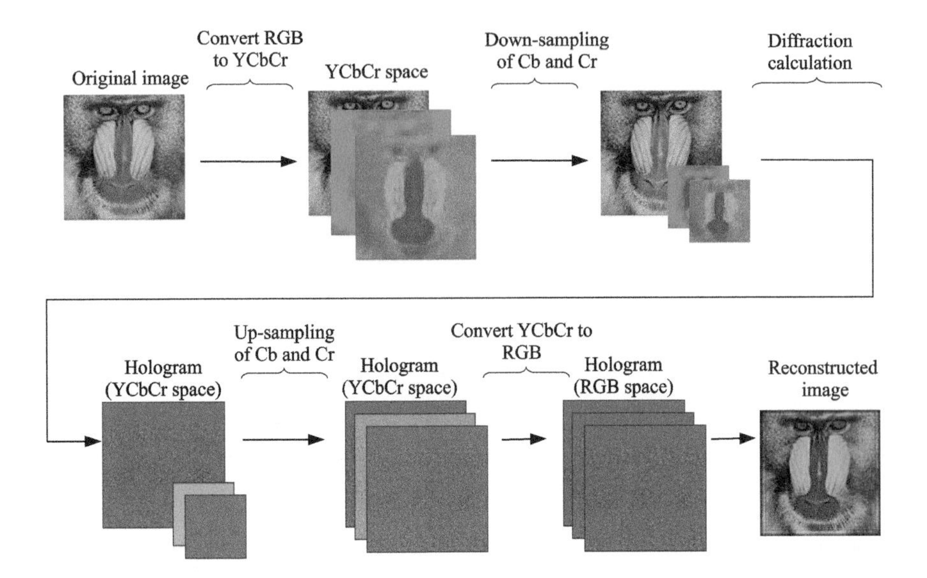

FIGURE 3.26 Color CGH calculation in YCbCr color space.

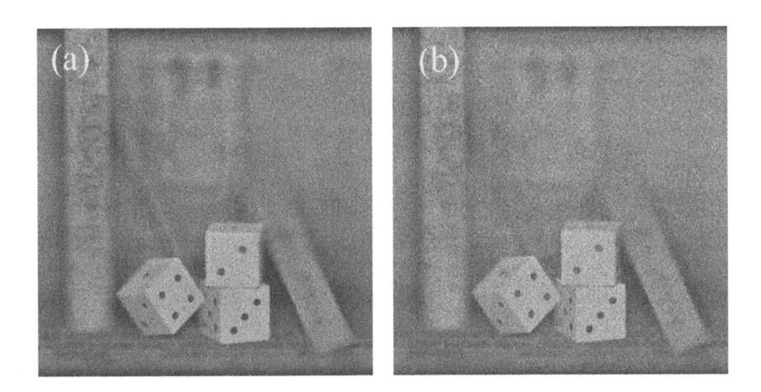

FIGURE 3.27 Reconstructed images. (a) Reconstruction from color CGH calculated in RGB color space and (b) reconstruction from color CGH calculated in YCbCr color space.

TABLE 3.2
Pros and cons of hologram calculations.

	Point Cloud	Polygon	Multiview image	RGB-D
Calculation cost	High	High	Low	Middle
Hidden surface removal and shading	Complicated	Complicated	Easy	Middle
Image quality	High	High	Average	High

the point cloud approach is more economical than the polygon approach unless the latter uses larger polygons.

Images reconstructed from cylindrical holograms have a 360° viewing angle; this makes cylindrical holographic display an appealing technique and has motivated several researchers to propose fast calculation algorithms for cylindrical holograms [89–93].

The computational complexity of an FFT-based diffraction is $O(N^2 \log N)$. Even when executed by FFT, a heavy computational load hinders the real-time calculation on a CPU of CGHs. However, computational complexity can be reduced to $O(N^2)$ by a Fresnel integral. In a previous study [94], the use of a Fresnel integral increased the execution speed by a factor of 9 relative to FFT-based diffraction, which was sufficient to observe a real-time reconstructed 3D movie at over 20 frames per second in a CPU. The use of a Fresnel integral has also been adopted in [95].

3.3.5 SUMMARY

Table 3.2 summarizes *pros and cons* of the above-mentioned hologram calculations. Unlike the multiview image approach, the point cloud and polygon approaches can reconstruct high-definition 3D images. Conversely, these methods exhibit high computational costs that can be reduced using the NLUT, WRP, tilted diffraction and analytical methods.

Unlike the point cloud and polygon methods, the multiview image approach has the merit of not requiring any hidden surface removal or shading processing because these are performed automatically by the computer graphics technique.

The RGB-D approach is superior to the multiview approach in its ability to reconstruct HD 3D images; however, in the RGB-D approach, it is difficult to reconstruct 3D images with a wide viewing angle because the RGB-D image is captured at a fixed position. It is likely that a hybrid approach combining the multiview image and RGB-D approaches would be a promising technique in terms of computational cost, image quality, and viewing angle.

Related algorithms have been thoroughly reviewed in Ref. [96–100].

4 Digital holography

Recently, hologram recording methods with an imaging device such as a CCD camera instead of a photographic dry plate are being actively conducted. Such an electronic hologram recording is called **digital holography**. The attraction of digital holography does not require the mechanical movement of optical elements and can observe an object light to be recorded three-dimensionally only by calculation. In addition, it is possible to simultaneously measure the amplitude and phase information of the object light. A digital holographic microscope that applies digital holography to a microscope has the feature of simultaneously recording the light transmittance and thickness information (phase information) of a microscopic object on a hologram by one shot. In this chapter, we first explain the principle of digital holography and introduce various application examples including digital holographic microscopy (DHM).

4.1 DIGITAL HOLOGRAPHY

Digital holography records holograms as digital data using imaging devices such as a **CCD** (**Charge-Coupled Device**) image sensor and a **CMOS** (**Complementary MOS**) image sensor. The attraction of digital holography lies in the ability to dynamically measure the amplitude and phase information of the object light at the same time.

A **digital holographic microscope** (**DHM**), which applies digital holography to the observation of a microscopic object, can measure the light transmittance and thickness information (phase information) of a microscopic object at the same time.

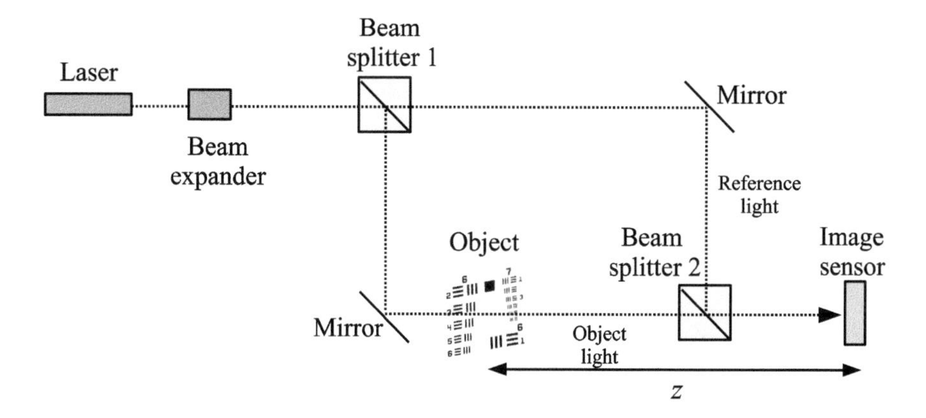

FIGURE 4.1 Example of a digital holographic microscope.

Figure 4.1 shows a general optical system of a DHM. The object is placed at the position z from the CCD. Since the beam diameter of the laser beam is narrow, it is enlarged with a **beam expander**. Then, it is separated into two beams by a beam splitter. As for the beam expander, the beam diameter can be expanded by combining two lenses.

One of the laser beams is irradiated to an object (such as a microorganism or a cell) to be recorded. This light is object light. The object light includes the amplitude (corresponding to transmittance) and phase information (corresponding to thickness information) of the object. The other laser light becomes the reference light. The reference light and the object light interfere with each other on the image sensor, and a hologram is recorded.

This process is described by a mathematical expression. Light waves can be described by amplitude and phase. Let $a_o(x_o, y_o)$ be the amplitude of the object and $\phi_o(x_o, y_o)$ be the phase of the object. The light waves on the surface where the object exists can be expressed as

$$u_o(x_o, y_o) = a_o(x_o, y_o) \exp(i\phi_o(x_o, y_o)). \tag{4.1}$$

When this object light propagates to the hologram, by using the operator of diffraction calculation (Section 2.5), the object light on the hologram plane can be written as

$$O(x, y) = \mathrm{Prop}_z[u_o(x_o, y_o)]. \tag{4.2}$$

When the reference light on the hologram plane is described as $R(x, y)$, the interference fringe $I(x, y)$ (hologram) of the object light and the reference light on the image sensor is described as

$$\begin{aligned} I(x, y) &= |O(x, y)|^2 + |R(x, y)|^2 + O(x, y)R^*(x, y) + O^*(x, y)R(x, y) \\ &= D(x, y) + O(x, y)R^*(x, y) + O^*(x, y)R(x, y) \end{aligned} \tag{4.3}$$

where $D(x, y) = |O(x, y)|^2 + |R(x, y)|^2$ is a direct light and $*$ denotes a complex conjugate.

Although the second term $O(x, y)R^*(x, y)$ is modulated by R^*, the object light is included in the second term, so that it can be seen that the amplitude and phase information of the object light is recorded in the hologram. However, in addition to the object light, unnecessary components of direct light ($D(x, y)$ in the first term) and conjugate light ($O^*(x, y)R(x, y)$ in the third term) are recorded.

In digital holography, a hologram recorded by the image sensor is transferred to a computer to reconstruct the object light by numerical calculation. For the reconstruction, the object light on a hologram can be numerically reconstructed by irradiating the same light (we call it **reconstruction light**) as the reference light at the time of recording to the hologram. The mathematical description that illuminates the reconstruction light to the hologram can be expressed by multiplying the hologram by the reconstruction light:

$$I(x, y) \times R(x, y) = R(x, y)D(x, y) + O(x, y) + O^*(x, y)R^2(x, y). \tag{4.4}$$

Since the second term is the recorded object light itself, it can be seen that both the amplitude and phase of object light can be reconstructed.

To calculate Eq. (4.4) on a computer, we simply pixel-wise multiply $R(x, y)$ with the hologram $I(x, y)$. Note that this calculation result is the light wave distribution on the hologram plane. Since the recoded object is actually located at a distance of $-z$ from the hologram plane, **inverse diffraction calculation (back propagation)** should be calculated from the result of Eq. (4.4) to obtain the light wave distribution at the position where the object was located. The reconstruction process is expressed as

$$\begin{aligned}\mathrm{Prop}_{-z}[I(x,y) \times R(x,y)] &= \mathrm{Prop}_{-z}[R(x,y)D(x,y)] + \\ &\quad \mathrm{Prop}_{-z}[O(x,y)] + \mathrm{Prop}_{-z}[O^*(x,y)R^2(x,y)].\end{aligned} \tag{4.5}$$

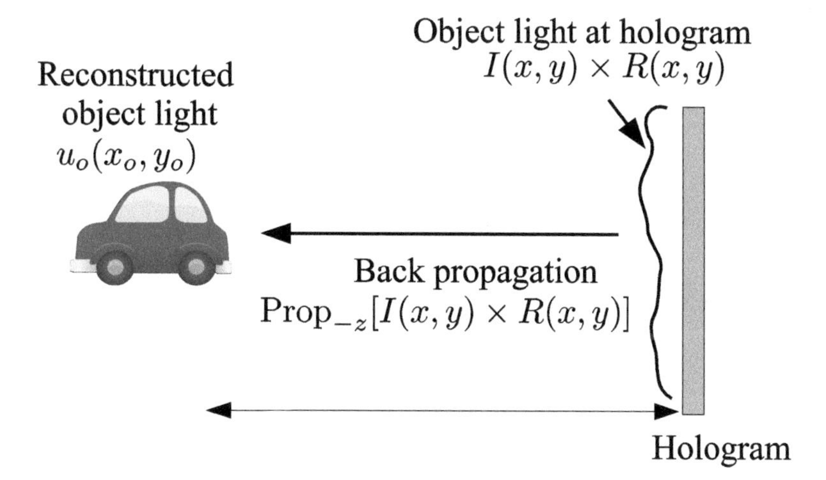

FIGURE 4.2 Inverse diffraction calculation.

Substituting Eq. (4.2) into Eq. (4.5), the second term of Eq. (4.5) is expressed as

$$\mathrm{Prop}_{-z}[O(x,y)] = \mathrm{Prop}_{-z}[\mathrm{Prop}_{z}[u_o(x_o, y_o)]] = u(x_o, y_o). \tag{4.6}$$

It can be seen that the object light recorded in the hologram can be recovered.

Since there is a relationship as shown in Figure 4.3, the amplitude of the reconstructed object light is calculated by

$$a_o(x_o, y_o) = \sqrt{\mathrm{Re}\{u_o(x_o, y_o)\}^2 + \mathrm{Im}\{u_o(x_o, y_o)\}^2}. \tag{4.7}$$

The phase is also calculated by

$$\phi_o(x_o, y_o) = \tan^{-1}\left(\frac{\mathrm{Im}\{u_o(x_o, y_o)\}}{\mathrm{Re}\{u_o(x_o, y_o)\}}\right). \tag{4.8}$$

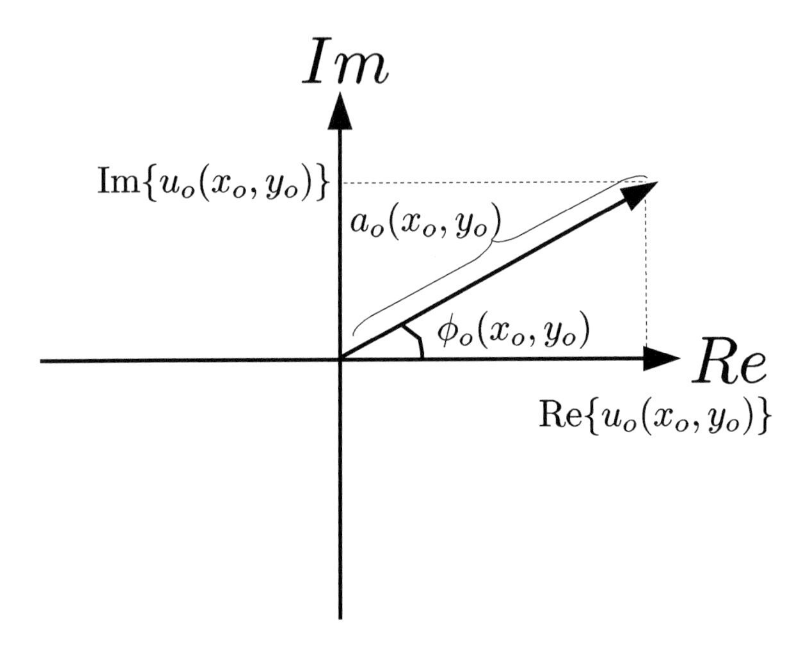

FIGURE 4.3 Amplitude and phase of object light.

The value range of $\phi_o(x_o, y_o)$ is $-\pi$ to π radian from the property of $\tan^{-1}$.

Consider the physical meaning of the phase obtained by Eq. (4.8). For example, when measuring a cell of Figure 4.4 with a DHM, a plane wave incident on the cell is modulated when it passes through the cell. The transmitted light that has passed through a thick part of the cell travels slowly so that a phase difference occurs in the wavefront of transmitted light as shown in Figure 4.4. Assuming that the refractive index distribution of the cell is $n(x, y, z)$, the phase difference of the object light measured by digital holography is calculated by

$$\phi_o(x_o, y_o) = \frac{2\pi}{\lambda} \int n(x, y, z) dz. \tag{4.9}$$

If $n(x, y, z)$ is known or the refractive index in the cell can be regarded as uniform, we can measure the cell thickness from $\phi_o(x, y)$.

Figure 4.5 shows an example of observing transmission and reflective objects using a DHM [2]. Figure 4.6(a) is a recorded hologram of human red blood cells, and Figures 4.6(b) and (c) are the amplitude and phase of the object light obtained by diffraction calculation from the hologram. The phase obtained by Eq. (4.8) is wrapped from $-\pi$ to π (white pixels in Figure 4.6(c) denote $-\pi$ and black pixels denote $+\pi$). Figure 4.5(d) shows an image obtained by performing **phase unwrapping** to convert the wrapped phase into an original thickness image, and Figure 4.5(e) shows a three-dimensional image of the human red blood cells from Figure 4.5(d).

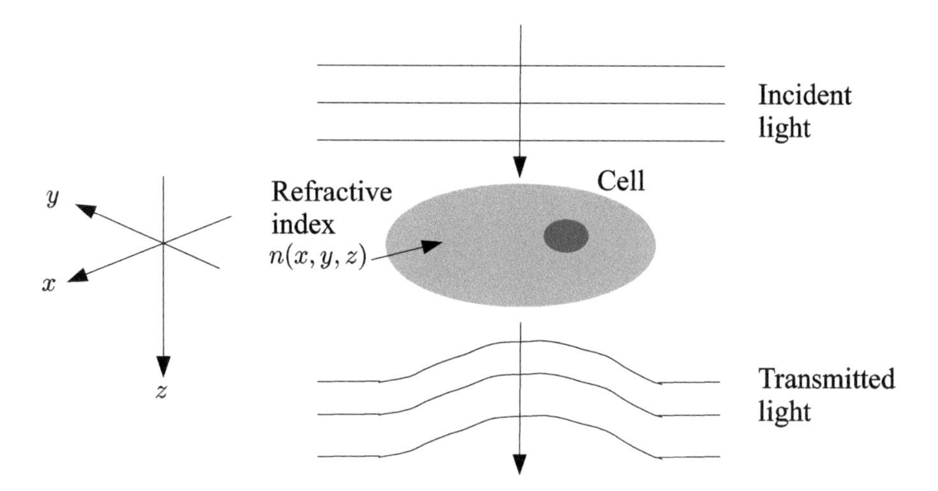

FIGURE 4.4 Example of phase measurement.

Not only can digital holography be used for a microscope, it can also perform three-dimensional measurement of a larger object. Reference [3] is an example of recording large objects. Figure 4.7 records three dice with digital holography. The left side of Figure 4.7 is an optical system that can capture a **lensless Fourier hologram**. By sequentially moving a single image sensor, a large-sized hologram can be captured. This is called **synthetic aperture digital holography**. The hologram records information when viewing the object light from various viewpoints. Once the hologram is captured, we can obtain a reconstructed image from various viewpoints simply by changing the parameters of the diffraction calculation as shown in Figures 4.7(a) - (d). This is a major feature of digital holography.

4.2 PROBLEMS OF DIGITAL HOLOGRAPHY

Although digital holography has the advantage of simultaneously acquiring the amplitude and phase of objects and a three-dimensional measurement without mechanical movement, the main drawbacks are as follows:

- Conjugate light and direct light are reconstructed in addition to object light (twin-image problem).
- Lack of a sampling interval for the hologram by the image sensor.
- Calculation time is required for diffraction calculation.

In Eq. (4.4), unnecessary lights (direct light and conjugate light) are reproduced. Depending on recording conditions (especially inline optical systems), these unnecessary lights overlap the object light and hinder the observation of the object light. This problem is called the **twin-image problem**.

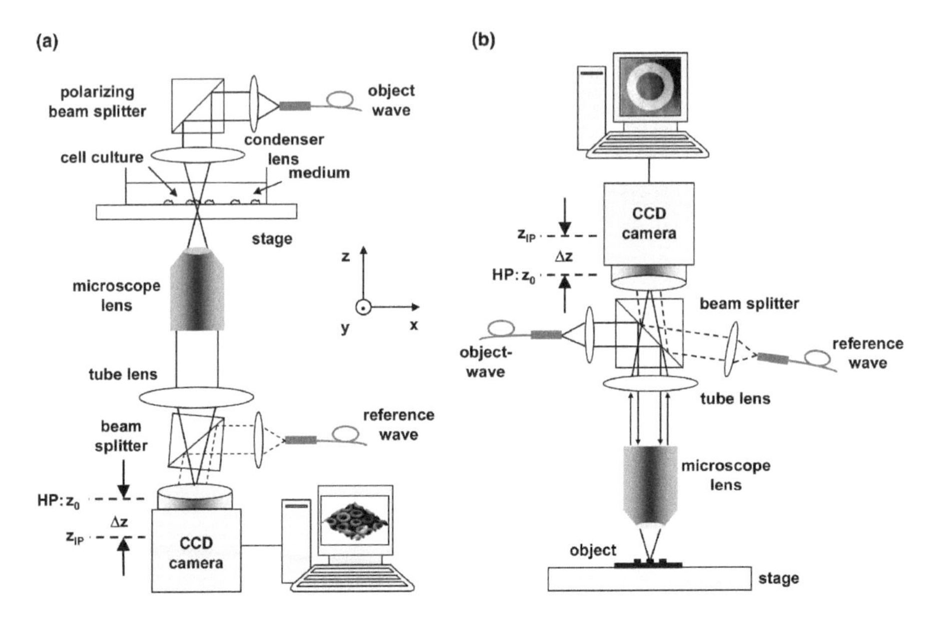

FIGURE 4.5 Example of a digital holographic microscope. (a) The optical system for transmission objects and (b) the optical system for reflective objects. Reprinted from Ref. [2] with permission, OSA.

In order to avoid this, one of the solutions is to adopt an **off-axis holography** that records a hologram using tilted reference light with an angle. Since the object light, the direct light, and the conjugate light are separated from each other, as the angle of the reference light is larger, it becomes easier to extract only the object light. However, if the angle of the reference light is increased, the interference fringe interval of the hologram becomes finer, making it difficult to capture the interference fringes with the image sensor (the second drawback). Therefore, unfortunately, the large angle of the reference light cannot be used in current CCD and CMOS sensors.

As shown in Figure 4.8, let us calculate the interference fringe interval d when the light emanating from one point of an object interferes with the reference light at $\mathbf{x} = (x, y)$ on the hologram plane. The wave number of the object light and reference light denote $k = 2\pi/\lambda$ (λ is the wavelength).

The object light emitted from a certain point of the object is expressed by

$$U_o = a_o \exp(i\mathbf{k_o} \cdot \mathbf{x}), \tag{4.10}$$

where $\mathbf{k_o} = (k_{o_x}, k_{o_y}) = (k \sin\theta_o, k \cos\theta_o)$ is the **wave vector** of the object light. The reference light is expressed by

$$U_r = a_r \exp(i\mathbf{k_r} \cdot \mathbf{x}), \tag{4.11}$$

where $\mathbf{k_r} = (k_{r_x}, k_{r_y}) = (-k \sin\theta_r, k \cos\theta_r)$ is the wave vector of the reference light.

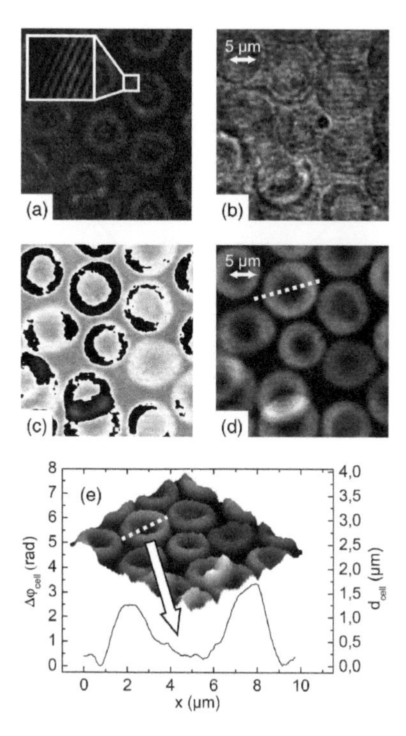

FIGURE 4.6 (a) Recorded hologram of human red blood cells, (b) reconstructed image (amplitude), (c) reconstructed image (phase), (d) phase unwrapped image, and (e) 3D representation of the phase distribution. Reprinted from Ref. [2] with permission, OSA.

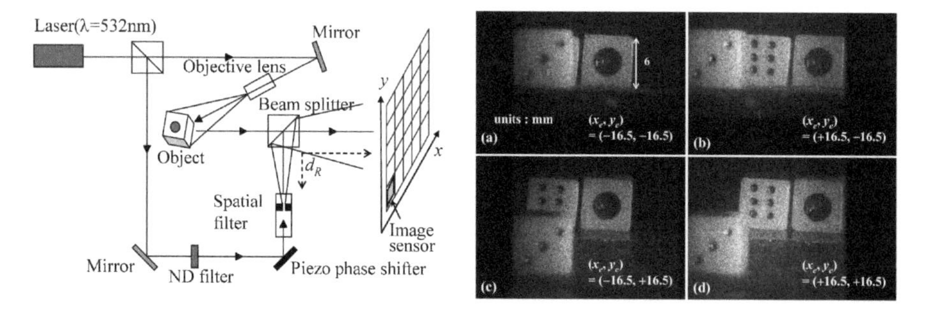

FIGURE 4.7 Three-dimensional image acquisition by digital holography. Reprinted from Ref. [3] with permission, OSA.

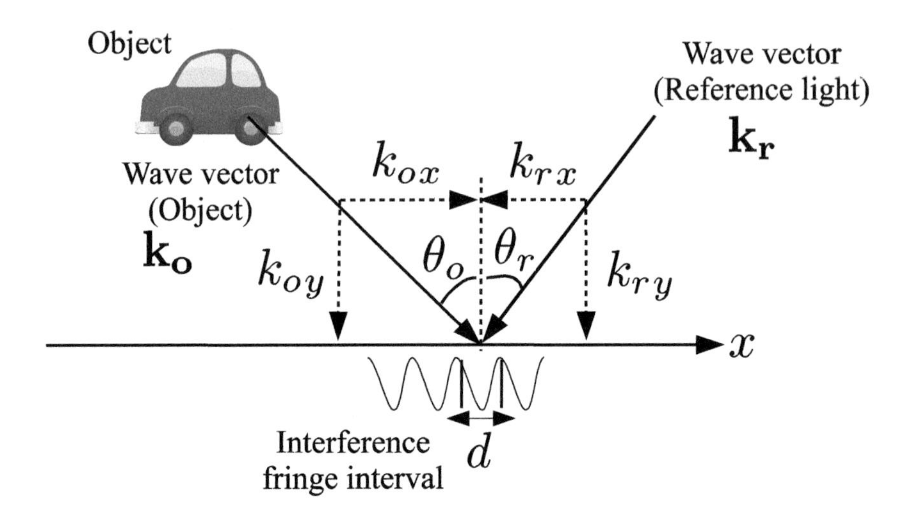

FIGURE 4.8 Interference fringe spacing of the hologram.

The interference pattern between the object and reference lights is expressed by

$$|U_o+U_r|^2 = |U_o|^2+|U_r|^2+U_oU_r^*+U_o^*U_r = a_o^2+a_r^2+2a_oa_r\cos((\mathbf{k_o}-\mathbf{k_r})\cdot\mathbf{x}). \tag{4.12}$$

Since only the phase of the interference fringe is important, unnecessary coefficients and terms are omitted as follows:

$$|U_o+U_r|^2 \propto \cos((\mathbf{k_o}-\mathbf{k_r})\cdot\mathbf{x}) = \cos(\phi), \tag{4.13}$$

where ϕ is defined by

$$\phi = k(\sin\theta_o+\sin\theta_r)x + k(\cos\theta_o-\cos\theta_r)y. \tag{4.14}$$

The spatial frequency along the x-axis is calculated by[i]

$$f = \frac{1}{2\pi}\left|\frac{d\phi}{dx}\right| = \frac{\sin\theta_o+\sin\theta_r}{\lambda}. \tag{4.15}$$

By calculating the reciprocal, the interference fringe interval d can be obtained by

$$d = \frac{\lambda}{\sin\theta_o+\sin\theta_r}. \tag{4.16}$$

[i]As mentioned in Section 2.6.4, the spatial frequency can be obtained by calculating the differential of the phase.

For example, under the condition of a red laser ($\lambda = 633$ nm), $\theta_o = 2°$ (about 0.035 radians), $\theta_r = 8°$ (about 0.14 radians), and d will be about 3.6 μm. When sampling this by each pixel of the image sensor, according to the sampling theorem, it is necessary to perform sampling at an interval finer than 1/2 of the interference fringe interval. Under this sampling condition, the pixel pitch of the image sensor is required to be finer than 1.8 μm.

In recent years, image sensors with such a pixel pitch are available on the market, but as the pixel pitch becomes finer, the imaging area of the image sensor also decreases, so the **field of view (FOV)** is decreased, which is the observable area of the object.

In digital holography, there are trade-off relationships among the twin-image problem, sampling problem, FOV, and resolution of reconstructed images, so we need to make a compromise. In the following, we will introduce representative optical systems of digital holography and methods to alleviate these problems.

4.3 INLINE DIGITAL HOLOGRAPHY

Figure 4.1 shows an example of an **inline digital holography** setup. Since the laser has a thin beam, we use a beam expander to make it an appropriate beam diameter to irradiate the entire object. In the beam splitter 1, the laser is split into two beams. The one is used for reference light and the other is irradiated on the object. The reference light and object light are synthesized by the beam splitter 2, and an interference fringe (hologram) is captured by the image sensor.

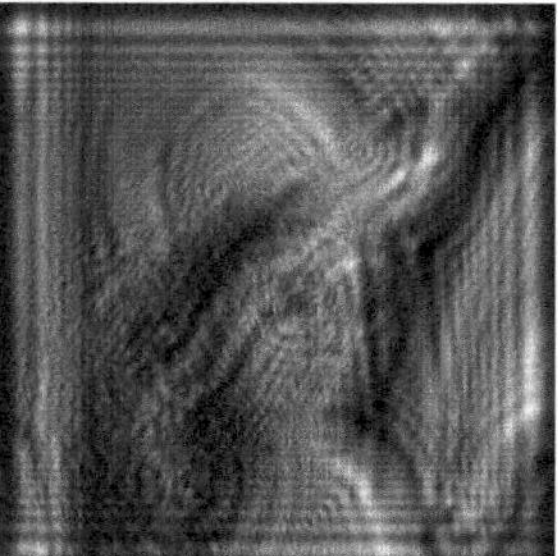

FIGURE 4.9 Reconstruction image from an inline hologram.

Figure 4.9 shows the simulation results of the inline setup. Figure 4.9 (left) is an object, and this object is irradiated with a plane wave with a wavelength of 633 nm. The object whose size is 2.56 mm $\times$ 2.56 mm is located at 0.1 m from the image sensor. A hologram recorded by the image sensor (256×256 pixels and the pixel pitch of 10 μm) is shown in Figure 4.9 (center). In order to obtain a reconstructed image from this hologram, diffraction calculation from the hologram to the position of the original object is performed. The reconstructed image is shown in Figure 4.9 (right). In the principle of the inline hologram, the direct light and conjugate light are superimposed on the reconstructed object light, which makes it difficult to observe. This is disadvantageous; however, since there is no angle of the reference light, it is advantageous that the interference fringe interval becomes gentler than the off-axis setup introduced in the next section.

4.4 OFF-AXIS DIGITAL HOLOGRAPHY

Figure 4.10 shows an example of **off-axis digital holography**. In an inline hologram, object light, direct light, and reference light completely overlap, and the image quality significantly deteriorates. In order to avoid the superimposition of these unnecessary lights, it is possible to use the off-axis optical system. By tilting the mirror on the reference light arm, the reference light can be obliquely incident and an off-axis hologram can be recorded by an image sensor.

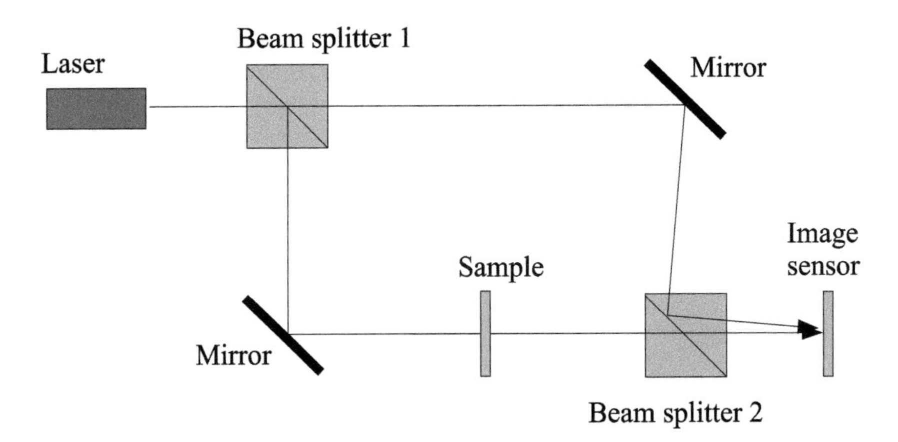

FIGURE 4.10 Off-axis digital holography.

The reference light is expressed by

$$R(x, y) = \exp(i(kx \sin\theta_x + ky \sin\theta_y)), \tag{4.17}$$

where θ_x and θ_y are angles from the optical axis. We assume the amplitude of 1, $k = 2\pi/\lambda$ is the wavenumber and λ is the wavelength. A hologram I is expressed by

$$I = |O + R|^2 = |O|^2 + |R|^2 + OR^* + O^*R \tag{4.18}$$

where $O(x, y)$ is the object light. Figure 4.11 (left) shows a hologram with the angles $\theta_x = 1°$ (about 0.017 rads) and $\theta_y = 0°$ of the reference light.

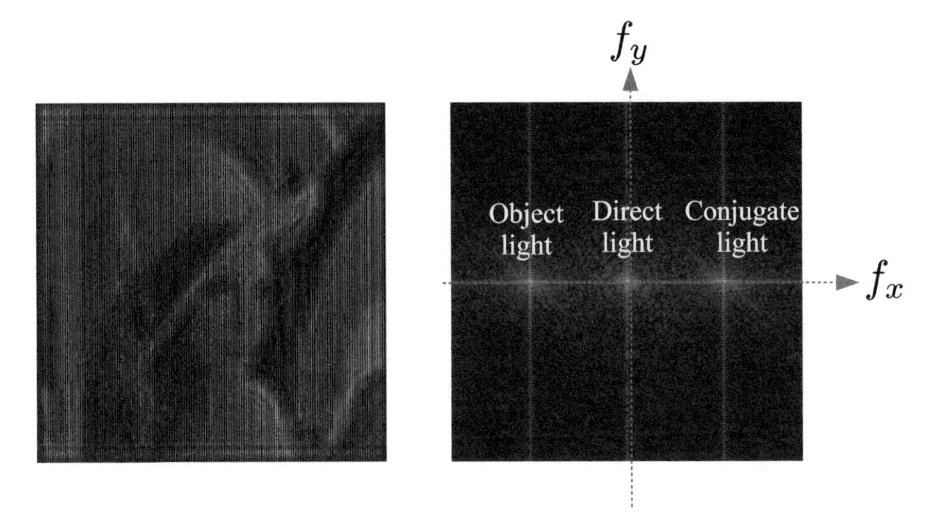

FIGURE 4.11 Off-axis hologram (left). The spectrum of the hologram (right).

Figure 4.11 (right) is the spectra of the hologram obtained by FFT. We consider how the spectra of the object light, conjugate light, and direct light are distributed. The Fourier transform of the hologram I is calculated by

$$\mathcal{F}\Big[I\Big] = \mathcal{F}\Big[|O|^2 + |R|^2\Big] + \mathcal{F}\Big[OR^*\Big] + \mathcal{F}\Big[O^*R\Big]. \tag{4.19}$$

The first term on the right side is the spectrum of the direct light, which is a low-frequency component. The second term on the right side is the spectrum of the object light. We pick up the second team as

$$\begin{aligned} \mathcal{F}\Big[O(x,y)R(x,y)^*\Big] &= \mathcal{F}\Big[O(x,y)\exp(-i(kx\sin\theta_x + ky\sin\theta_y))\Big] \\ &= \tilde{O}(f_x + \frac{\sin\theta_x}{\lambda}, f_y + \frac{\sin\theta_y}{\lambda}), \end{aligned} \tag{4.20}$$

where $\tilde{O}$ is the spectrum of the object light O. The **Fourier shift theorem** described in Chapter 7 was used for this formula transformation.

In the Fourier domain, the original spectrum of the object light $\tilde{O}(f_x, f_y)$ is shifted to $(-\frac{\sin\theta_x}{\lambda}, -\frac{\sin\theta_y}{\lambda})$. Likewise, the third term on the right side of Eq. (4.19) is the spectrum of the conjugate light. We pick up the third team as

$$\begin{aligned} \mathcal{F}\Big[O^*(x,y)R(x,y)\Big] &= \mathcal{F}\Big[O^*(x,y)\exp(i(kx\sin\theta_x + ky\sin\theta_y))\Big] \\ &= \tilde{O}^*(f_x - \frac{\sin\theta_x}{\lambda}, f_y - \frac{\sin\theta_y}{\lambda}). \end{aligned} \tag{4.21}$$

In the Fourier domain, the original spectrum of the conjugate light $\tilde{O}(f_x, f_y)$ is shifted to $(+\frac{\sin\theta_x}{\lambda}, +\frac{\sin\theta_y}{\lambda})$.

Figure 4.12 shows the spectrum of the hologram when the reference light angles θ_x and θ_y are changed. In the case of the inline hologram ($\theta_x = 0°$, $\theta_y = 0°$), the spectrum of the direct light, conjugate light, and object light are completely overlapped. On the other hand, in the case of the off-axis hologram ($\theta_x = 1°$, $\theta_y = 0°$), it can be confirmed that the three spectra are separated in the horizontal direction. From the left, they are the object light, direct light, and conjugate light components. In the case of ($\theta_x = 0.5°$, $\theta_y = 0.5°$), the object light and conjugate light components are arranged diagonally. When the reference light angle is further doubled ($\theta_x = 1°, \theta_y = 1°$), the positions of the spectrum of the object light and of the conjugate light also change twice as much.

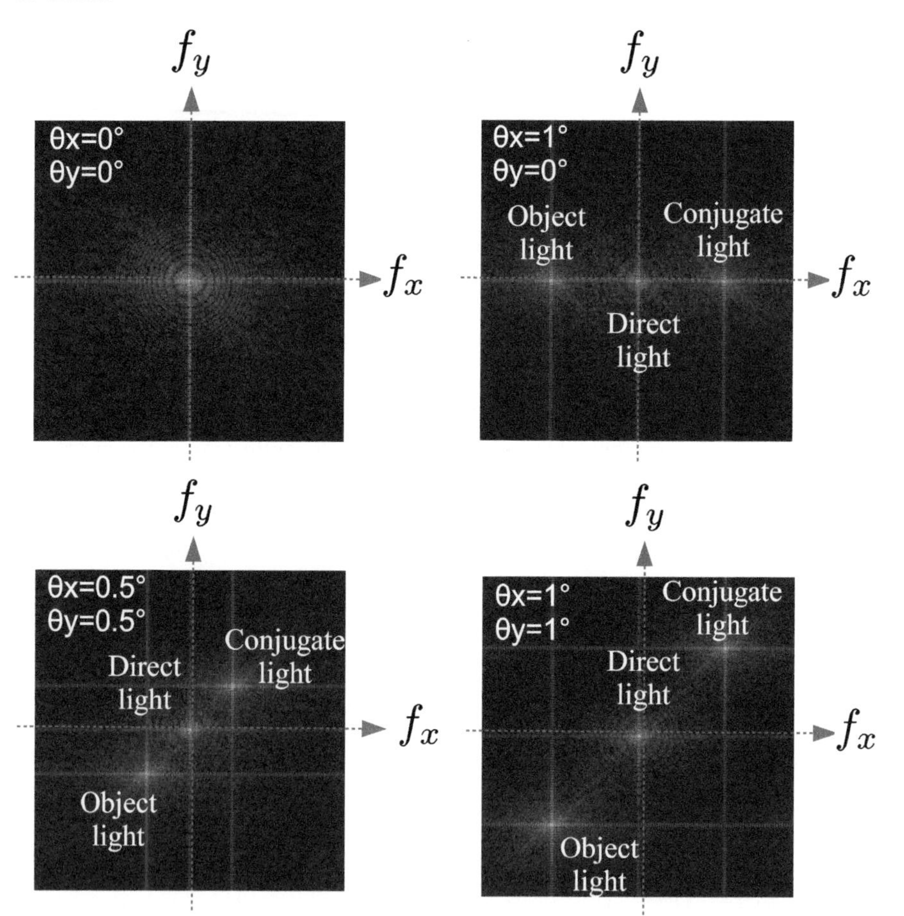

FIGURE 4.12 Spectra of the inline hologram and off-axis hologram.

Since the spectrum of the object light, the conjugate light, and the direct light are clearly separated in the frequency domain, it is easy to extract only the object light [101,102]. When the reference light angle is ($\theta_x = 1°$, $\theta_y = 1°$), only the object light spectrum is extracted (filtered) from Figure 4.13 (left). Next, as shown in Figure 4.13 (center), the extracted part is moved to the center of the frequency domain and **zero padding** is performed. After that, when performing the inverse FFT, only the object light can be reconstructed as shown in Figure 4.13 (right). Note that the object light is on the hologram plane; therefore, it is necessary to calculate the inverse diffraction to the position where the original object was present as shown in Figure 4.2.

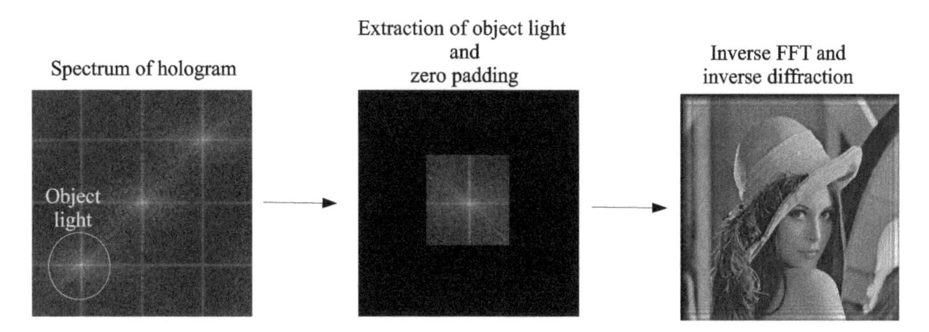

FIGURE 4.13 Reconstruction for the off-axis hologram.

4.5 GABOR DIGITAL HOLOGRAPHY

Figure 4.14 shows an optical setup of **Gabor holography**. The optical system used by D. Gabor, inventor of holography, uses a simple optical system that does not separate a laser beam into two beams, like inline and off-axis setups. In some literature, it is sometimes called an inline holography, but in this book it is called Gabor holography.

In Gabor holography, it is assumed that an object is small. We illuminate the object with a plane wave $r(x_o, y_o)$. The part of the object on which the plane wave impinges becomes the object light. The object light is expressed by $u_o(x_o, y_o)$. Light passing through a portion where the object is not present is regarded as the reference light. Therefore, the object plane can be described as $1 - u_o(x_o, y_o)$ where 1 means an area in the absence of the object and u_o is assumed to be an absorbing object.

The complex amplitude on the hologram plane is expressed by

$$\begin{aligned} U(x,y) &= \mathrm{Prop}_z[r(x_o,y_o)(1-u_o(x_o,y_o))] && (4.22)\\ &= \mathrm{Prop}_z[r(x_o,y_o)] - \mathrm{Prop}_z[r(x_o,y_o)u_o(x_o,y_o)] && (4.23)\\ &= R(x,y) - O(x,y), && (4.24)\end{aligned}$$

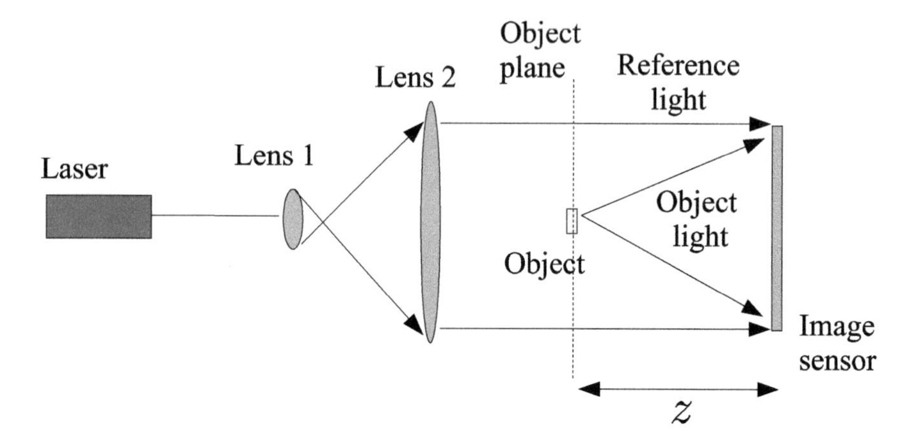

FIGURE 4.14 Gabor holography.

where we define $R(x, y)$ and $O(x, y)$ as follows:

$$R(x, y) = \text{Prop}_z[r(x_o, y_o)] \tag{4.25}$$

$$O(x, y) = \text{Prop}_z[r(x_o, y_o)u_o(x_o, y_o)]. \tag{4.26}$$

The hologram is expressed by

$$U(x, y) = |R(x, y)|^2 + |O(x, y)|^2 - O(x, y)R^*(x, y) - O^*(x, y)R(x, y). \tag{4.27}$$

Assuming the reference light of $R(x, y) = 1$, the reconstruction of the hologram is simply expressed by

$$\begin{aligned} \text{Prop}_{-z}[I(x, y)] &= \text{Prop}_{-z}[1 + |O(x, y)|^2] \\ &\quad - \text{Prop}_{-z}[O(x, y)] - \text{Prop}_{-z}[O^*(x, y)]. \end{aligned} \tag{4.28}$$

The first term is direct light, and the second and third terms are object light and conjugate light.

Using the property of the diffraction operator[ii] $\text{Prop}_z[\cdot]$, the second term, is expressed as

$$\begin{aligned} \text{Prop}_{-z}[O(x, y)] &= \text{Prop}_{-z}[\text{Prop}_z[r(x_o, y_o)u_o(x_o, y_o)]] \\ &= r(x_o, y_o)u_o(x_o, y_o). \end{aligned} \tag{4.29}$$

The third term is also expressed as

$$\begin{aligned} \text{Prop}_{-z}[O^*(x, y)] &= \text{Prop}_{-z}[\text{Prop}_{-z}[r^*(x_o, y_o)u_o^*(x_o, y_o)]] \\ &= \text{Prop}_{-2z}[r^*(x_o, y_o)u_o^*(x_o, y_o)]. \end{aligned} \tag{4.30}$$

Since the conjugate light propagates by the distance of $-2z$, this light blurs over the entire reconstructed image.

[ii]See Section 2.5.

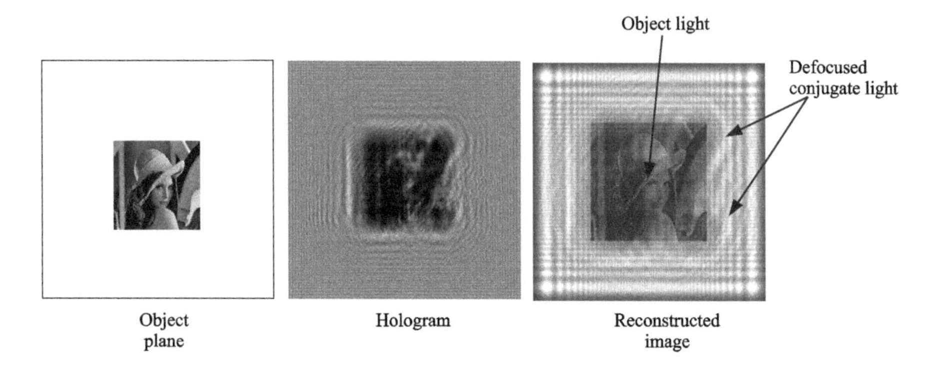

FIGURE 4.15 Reconstruction image from a Gabor hologram.

Figure 4.15 shows a reconstruction image from a Gabor hologram. The figure on the left represents the object on the object plane.[iii] The center of the figure is the Gabor hologram, and its reconstructed image is shown in the right side of the figure. The conjugate light is blurred in the whole reconstructed image.

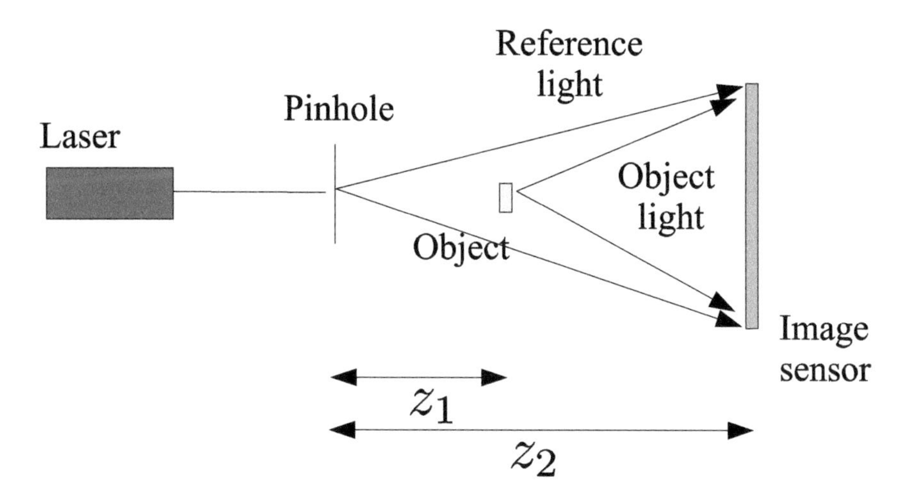

FIGURE 4.16 Lensless Gabor digital holography.

As shown in Figure 4.16, by using a pinhole, it is possible to construct a completely lensless optical system. This optical system has the feature that the observation magnification of a reconstruction image can be changed by

[iii]The white pixels transmit the incident light. The dark pixels attenuate the incident light.

the arrangement of the pinhole, the object, and the image sensor. The magnification M is geometrically determined by

$$M = \frac{z_2}{z_1}. \tag{4.31}$$

In addition, when using the optical system with the plane wave as the reference light (Figure 4.14), capturing a large hologram requires a large lens. In contrast, in Figure 4.16, there is an advantage that it is not necessary to prepare a large lens. A gigapixel DHM introduced in the latter uses this feature.

4.5.1 GABOR COLOR DIGITAL HOLOGRAPHIC MICROSCOPY

Two examples using Gabor digital holographic microscopy are introduced. Figure 4.17 shows Gabor color digital holographic microscopy [4].

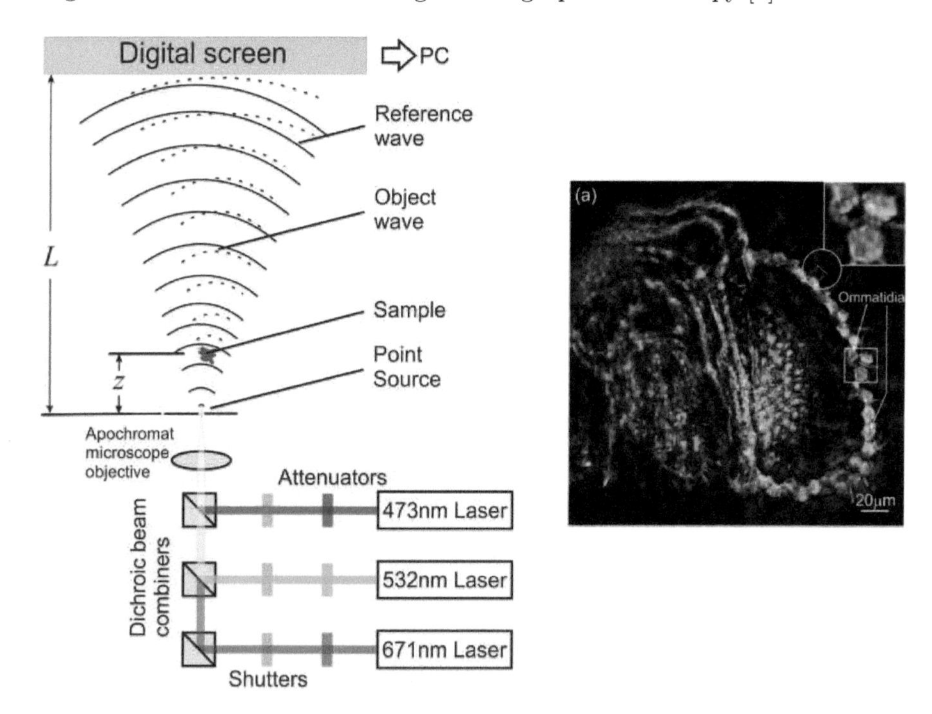

FIGURE 4.17 Gabor-type color digital holographic microscopy. Reprinted from Ref. [4] with permission, OSA.

The left side of the figure is an optical system, and color imaging is performed by introducing RGB laser light sources into a Gabor-type optical system. The right side of the figure is the color reconstructed image of a fruit fly eye (compound eye), and it is possible to observe the monocular eyes individually. In general, since an object has wavelength-dependent absorption, it is sometimes difficult to observe a certain part of the object at one wavelength, but more detailed observation can be performed by increasing the number of the wavelength for the reference lights.

4.5.2 GIGAPIXEL DIGITAL HOLOGRAPHIC MICROSCOPY

As another example, a **gigapixel digital holographic microscope** using a scanner [103] is introduced. DHMs have a problem that the hologram size to be captured is limited by the imaging device size. If a hologram to be captured has a large area, it is possible to observe the reconstructed image in a wide area; however, even for an imaging device with a large size, the size is generally limited to several square centimeters. Ultra-large imaging devices used for astronomical applications have been developed, but they are very expensive. In a general method of capturing a hologram with a large area using an existing imaging device, the hologram is acquired by sequentially moving the imaging device horizontally and vertically. Such a method is called **synthetic aperture digital holography**. In order to more conveniently capture a large-area hologram, there is a method of using a commercially available flatbed scanner for an imaging device.

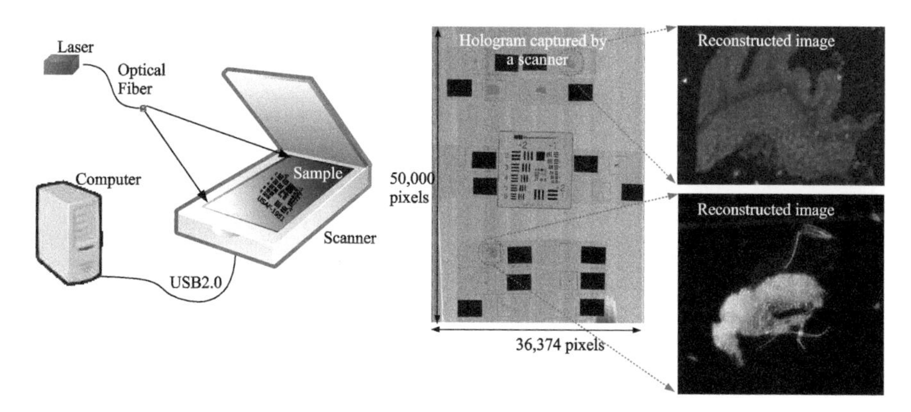

FIGURE 4.18 Gigapixel digital holography using a consumer scanner.

Figure 4.18 (left) is an optical system using a flat head scanner. We place an object directly on the glass surface of the scanner, then record a Gabor hologram by illuminating spherical light from an optical fiber output. The advantage of the Gabor-type optical system is that it can irradiate a wide area without using a special optical system. This scanner can directly scan a hologram with a one-dimensional imaging device which has the same width as an A4-sized document.

Figure 4.18 (center) is a hologram captured by actually placing multiple objects on the glass surface of the scanner. The number of pixels in the hologram is about $50{,}000 \times 36{,}000$ (about 1.8 gigapixels) and 4,800 **dots per inch (dpi)**. The sampling interval of 4,800 dpi is about 5.3 μm.

A reconstructed image from the A4-sized hologram can be obtained. A part of the reconstructed image from this hologram is shown in Figure 4.18 (right). The reconstructed images are the stomach wall of an animal (above) and a Drosophila (below).

4.6 PHASE-SHIFTING DIGITAL HOLOGRAPHY

In inline digital holography, the interference fringe interval becomes gentle as compared with off-axis holography, so the resolution power of the reconstructed image can be increased. However, there is a problem that the conjugate light and direct light are superimposed on the object light. On the other hand, in off-axis digital holography, only the object light can be reproduced by filtering. However, due to the filtering, the high-frequency component of the object light is lost and the resolution power becomes worse than in inline digital holography.

Phase-shifting digital holography is a technique that can only reproduce object light while maintaining the advantages of the inline type. Figure 4.19 shows the optical system of phase-shifting digital holography, which is almost the same as the optical system of inline holography. The only difference is to incorporate a device which can finely adjust the phase of the reference light.

Figure 4.19 precisely controls the phase of the reference light on nanometer order by using a **piezo element.**[iv]

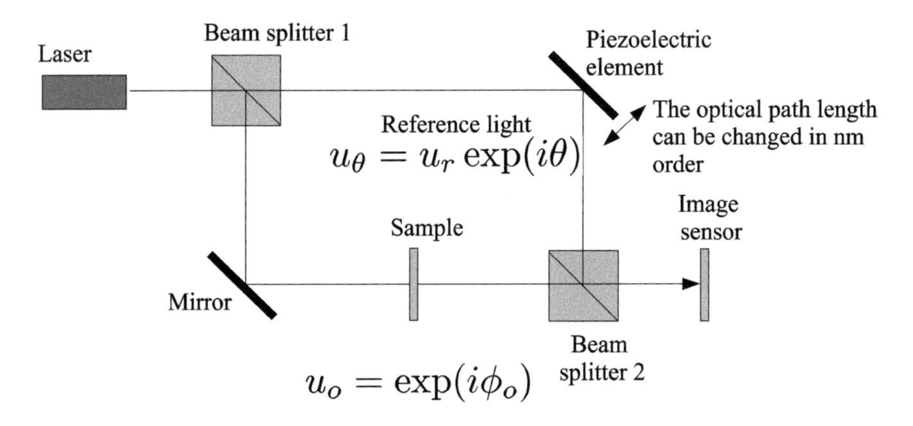

FIGURE 4.19 Phase-shifting digital holography.

Here we consider an object with the amplitude and phase distribution of

[iv]Alternatively, a phase plate capable of accurately delaying the phase of an incident light can be used.

Figure 4.20.[v] Let us see that this amplitude and phase can be restored by the phase- shifting digital holography.

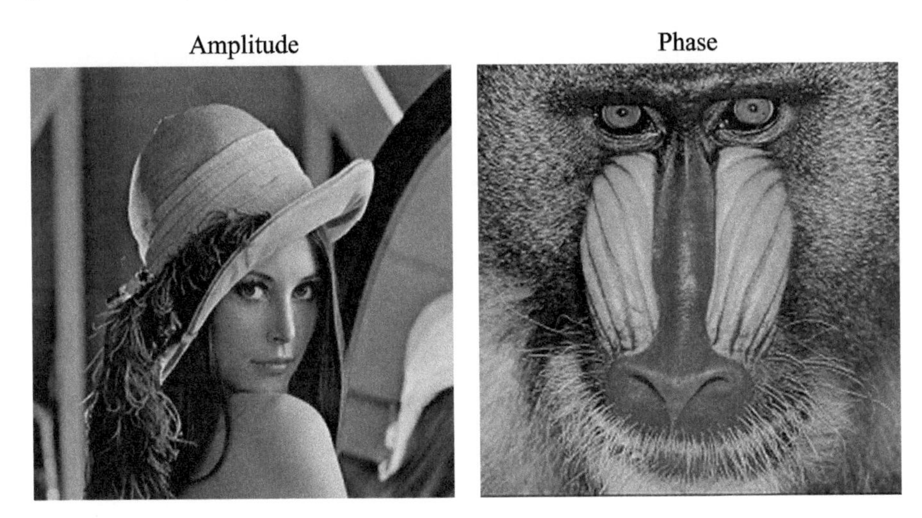

FIGURE 4.20 Original object. The left is the amplitude component. The right is the phase component.

We denote the object light on the hologram plane as U_o and the reference light as $U_r \exp(i\theta)$. The inline hologram I_θ, where the phase of the reference light is shifted by θ, can be written as

$$\begin{aligned} I_\theta &= |U_0 + U_r \exp(i\theta)|^2 \\ &= |U_o|^2 + |U_r|^2 + U_o U_r^* \exp(-i\theta) + U_o^* U_r \exp(i\theta). \end{aligned} \tag{4.32}$$

The phase-shifting digital holography records four inline holograms while shifting the phases of the reference light to 0, $\pi/2(=\lambda/4)$, $\pi(=\lambda/2)$ and $3\pi/2(=3\lambda/4)$. For example, in the case of a red laser (the wavelength λ is 633 nm), in order to shift the phase of the reference light by $\pi/2$ steps, the mirror attached to the piezo elements shifted by approximately 633 nm/4 $\approx$ 158 nm.

The inline holograms are expressed as

$$\begin{aligned} I_0 &= |U_o|^2 + |U_r|^2 + U_o U_r^* + U_o^* U_r, \\ I_{\frac{\pi}{2}} &= |U_o|^2 + |U_r|^2 - iU_o U_r^* + iU_o^* U_r, \\ I_\pi &= |U_o|^2 + |U_r|^2 - U_o U_r^* - U_o^* U_r, \\ I_{\frac{3}{2}\pi} &= |U_o|^2 + |U_r|^2 + iU_o U_r^* - iU_o^* U_r. \end{aligned} \tag{4.33}$$

Figure 4.21 shows the phase-shifted inline holograms.

[v]The left shows the amplitude distribution in 256 gradations, the right shows the phase distribution ($\pi \sim -\pi$) in 256 gradations.

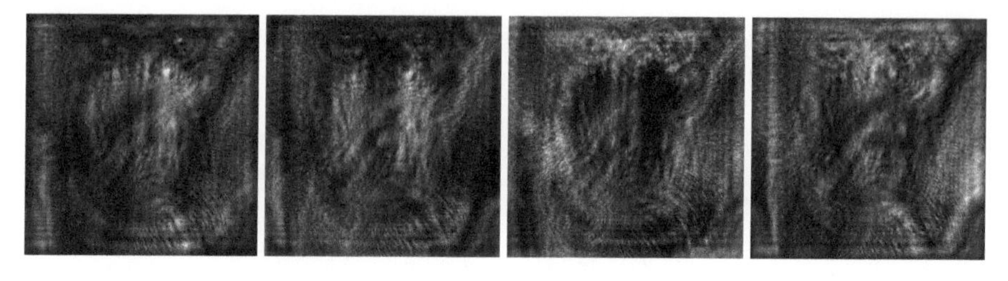

FIGURE 4.21 Phase-shifted holograms. From the left, the phase shifts are 0, $\pi/2(=\lambda/4)$, $\pi(=\lambda/2)$ and $3\pi/2(=3\lambda/4)$.

From Eq. (4.33), the following equation is obtained:

$$\begin{aligned}(I_0 - I_\pi) + i(I_{\frac{\pi}{2}} - I_{\frac{3}{2}\pi}) &= (2U_oU_r^* + 2U_o^*U_r) + (2U_oU_r^* - 2U_o^*U_r) \\ &= 4U_oU_r^*. \end{aligned} \tag{4.34}$$

Finally, the object light can be calculated by

$$U_o = \frac{1}{4U_r^*}((I_0 - I_\pi) + i(I_{\frac{\pi}{2}} - I_{\frac{3}{2}\pi})). \tag{4.35}$$

When the amplitude of the reference light can be assumed to be constant (e.g., 1), the coefficient can be omitted, so it can be further simplified to

$$U_o = ((I_0 - I_\pi) + i(I_{\frac{\pi}{2}} - I_{\frac{3}{2}\pi})). \tag{4.36}$$

Since U_o is the object light on the image sensor. When observing the reconstructed image, it is necessary to perform inverse diffraction calculation to the position where the object originally existed.

Figure 4.22 shows the simulation result of **4-step phase-shifting digital holography**. The left figure shows the result where the reconstruction is performed from a single inline hologram as a comparison. It can be seen that the direct light and conjugate light are superimposed in addition to object light. The right figure is a reconstructed image of the 4 step phase-shifting digital holography, showing that both the amplitude and phase of object light can be restored.

Figure 4.23 is the result of optical experiments. The object is a die. It is a reconstructed image obtained by using the phase-shifting method from four inline holograms. The left figure is a reconstructed image obtained using a single inline hologram. The conjugate light and direct light are superimposed on the object light. The right figure shows that only the object light can be reconstructed by using the phase-shifting method.

4.6.1 3-STEP PHASE-SHIFTING DIGITAL HOLOGRAPHY

Phase-shifting digital holography can reproduce only an object light. However, because it is necessary to record a plurality of holograms, it takes time to

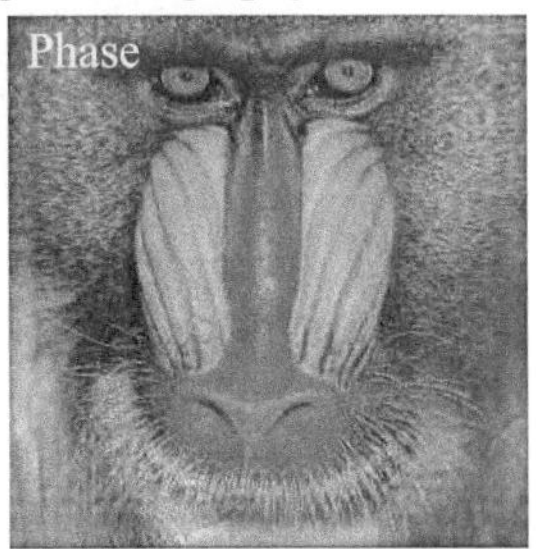

FIGURE 4.22 Reconstructed images. The left is the reconstructed image from a single inline hologram. The right images are reconstructed by the phase-shifting digital holography.

record, which is a disadvantage. Therefore, it is desirable for the number of hologram recordings to be small.

In **3-step phase-shifting digital holography** [104], three inline holograms are recorded by shifting the phases of the reference light of 0, $\pi/2$, and π. The formula is expressed as follows:

$$\begin{aligned}
I_0 &= |U_o|^2 + |U_r|^2 + U_o U_r^* + U_o^* U_r, \\
I_{\frac{\pi}{2}} &= |U_o|^2 + |U_r|^2 - iU_o U_r^* + iU_o^* U_r, \\
I_\pi &= |U_o|^2 + |U_r|^2 - U_o U_r^* - U_o^* U_r.
\end{aligned} \tag{4.37}$$

From these equations, we obtain the following equation:

$$(I_0 - I_{\frac{\pi}{2}}) + i(I_{\frac{\pi}{2}} - I_\pi) = 2(1+i)U_o U_r^*. \tag{4.38}$$

The three inline holograms and reference light U_r are known, so the object light can be calculated by

$$U_o = \frac{1-i}{4U_r^*}\left((I_0 - I_{\frac{\pi}{2}}) + i(I_{\frac{\pi}{2}} - I_\pi)\right). \tag{4.39}$$

Figure 4.24 shows the amplitude and phase of the reconstructed image with the three-step phase-shifting digital holography. For comparison, the reconstructed image of 4-step phase-shifting digital holography is also shown. In principle, the three-step and four-step digital holography have the same reconstructed image, but in an actual experiment, the three-step phase shifting digital holography is susceptible to noise since the number of steps is small.

4.6.2 2-STEP PHASE-SHIFTING DIGITAL HOLOGRAPHY

2-step phase-shifting digital holography [105] records two inline holo-

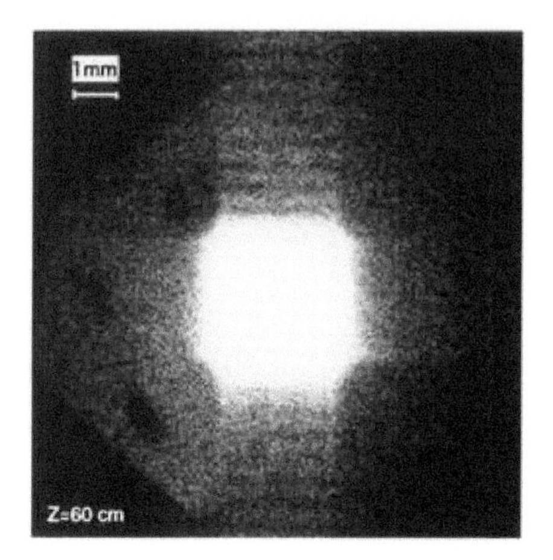

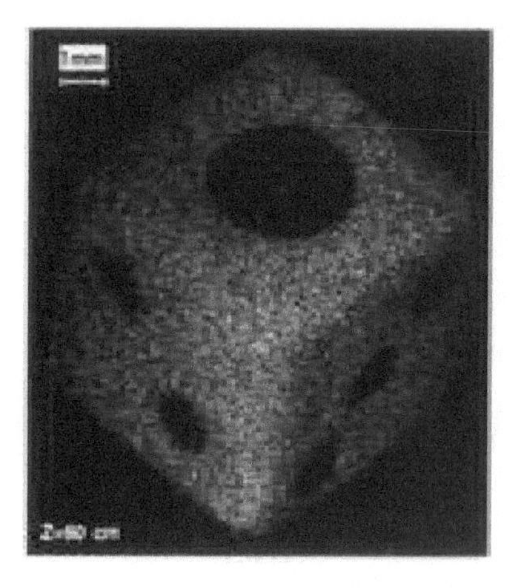

FIGURE 4.23 Experimental results. The left image is obtained by inline digital holography. The right image is obtained by phase-shifting digital holography. Reprinted from Ref. [5] with permission, OSA.

grams by shifting the phase of reference light to 0, θ.[vi] The equation is expressed as

$$\begin{aligned} I_0 &= |U_o|^2 + |U_r|^2 + U_o U_r^* + U_o^* U_r, \\ I_\theta &= |U_o|^2 + |U_r|^2 + U_o U_r^* \exp(-i\theta) + U_o^* U_r \exp(i\theta). \end{aligned} \tag{4.40}$$

Unlike the previous phase-shifting methods, the 2-step phase-shifting method requires the intensity distribution $|U_o|^2$ of object light and the complex amplitudes U_r (and $|U_r|^2$ calculated from U_r) of the reference light as known information.

Using the known information $|U_o|^2$ and $|U_r|^2$, we calculate

$$I_0 - |U_o|^2 - |U_r|^2. \tag{4.41}$$

Similarly, using the known information $|U_o|^2$, $|U_r|^2$ and $\exp(-i\theta)$, we calculate

$$\exp(-i\theta)(I_\theta - |U_o|^2 - |U_r|^2). \tag{4.42}$$

Subtracting Eq. (4.42) from Eq. (4.41) results in

$$\begin{aligned} &(I_0 - |U_o|^2 - |U_r|^2) - \exp(-i\theta)(I_\theta - |U_o|^2 - |U_r|^2) \\ &= U_o U_r^* (1 - \exp(-i2\theta)). \end{aligned} \tag{4.43}$$

Eventually, the object light can be calculated by

$$U_o = \frac{I_0 - |U_o|^2 - |U_r|^2 - \exp(-i\theta)(I_\theta - |U_o|^2 - |U_r|^2)}{U_r^*(1 - \exp(-i2\theta))}. \tag{4.44}$$

[vi]Other two-step phase-shifting digital holography techniques have also been proposed.

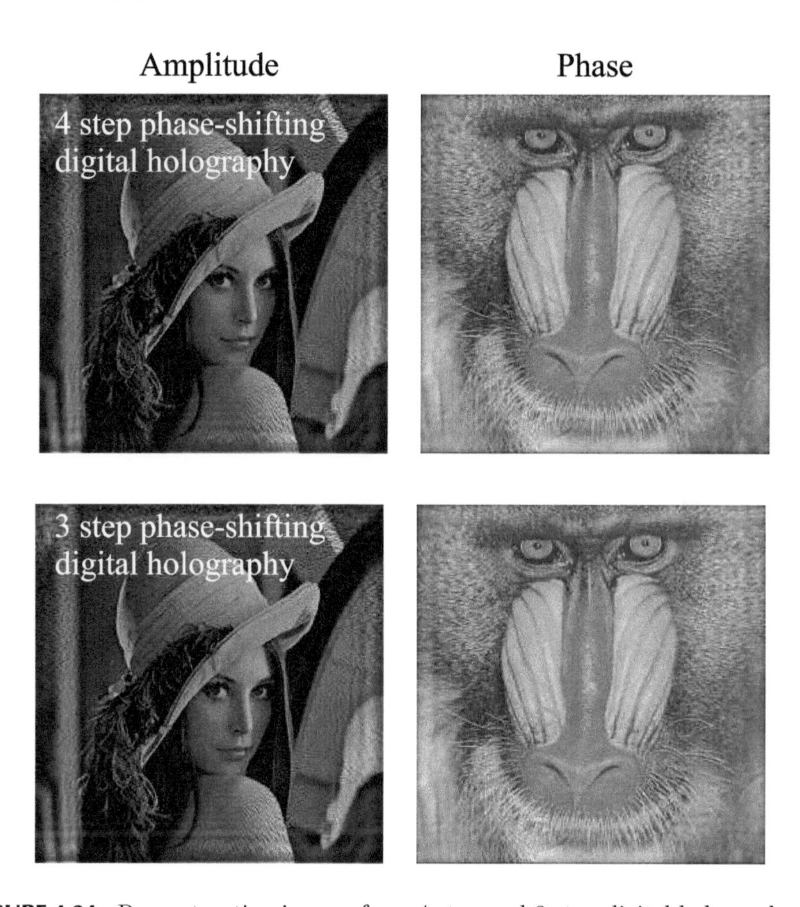

FIGURE 4.24 Reconstruction images from 4-step and 3-step digital holography.

4.6.3 1-STEP PHASE-SHIFTING DIGITAL HOLOGRAPHY

Although phase-shifting digital holography has an advantage that only object light can be reproduced, there is a disadvantage that multiple hologram recordings are necessary, so high-speed imaging is difficult to perform. For high-speed recording, it is desirable to realize phase-shifting digital holography with one shot (that is, **1-step phase-shifting digital holography**). Several methods have been proposed for one-step phase-shifting digital holography. Among these methods, here, we introduce **parallel phase-shifting digital holography** [6].

Figure 4.25(a) shows the optical setup of the parallel phase-shifting digital holography. Basically, it is an inline-type optical system, but a phase-shifting array device is placed on the reference arm. The phase-shifting array device is shown in Figure 4.25(b). This array device is designed so the incident planar wave simultaneously causes phase delays of $0, \pi/2, \pi$, and $3\pi/2$. Such a phase-

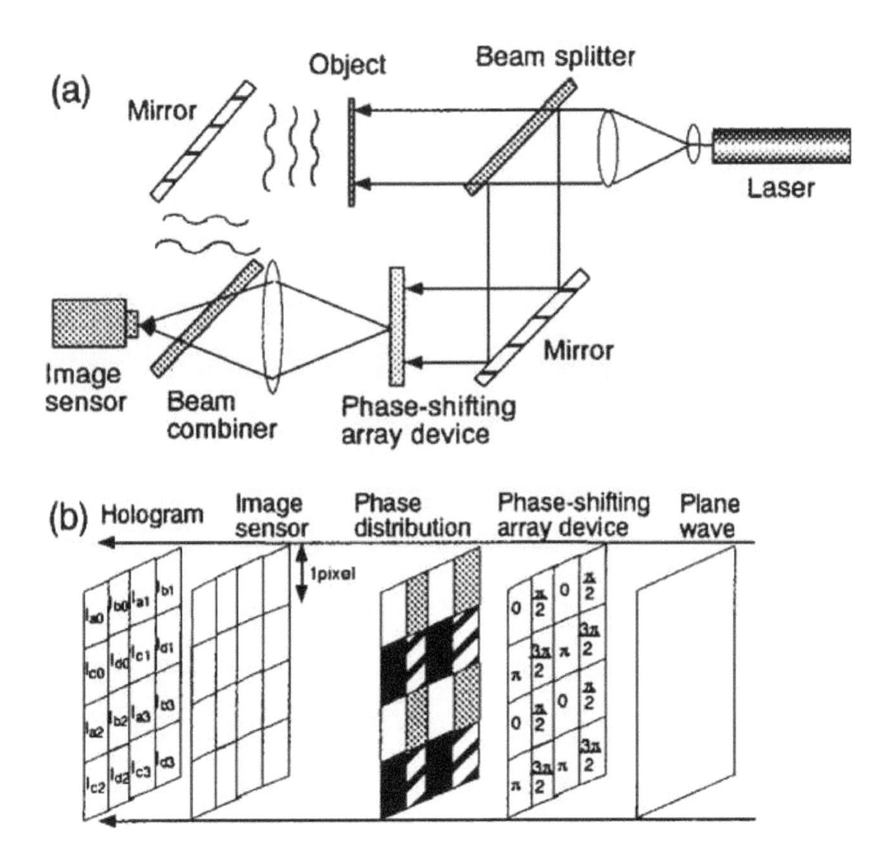

FIGURE 4.25 Parallel phase-shifting digital holography. Reproduced from [6] , with the permission of AIP Publishing.

shifting array device can be realized by, for example, processing the thickness of the glass plate so as to obtain the required phase shifts.

The phase-shifted reference light is associated with each pixel of the image sensor through the lens. By interfering this reference light with the object light on the image sensor, it is possible to simultaneously record four phase-shifted holograms of $0, \pi/2, \pi$, and $3\pi/2$.

Figure 4.26 shows the reconstruction algorithm. One pixel of phase-shifted holograms of $0, \pi/2, \pi$, and $3\pi/2$ are recorded on 2×2 pixels of the image sensor. If pixels with the same phase-shifting amount are extracted, Figure 4.26(b) is obtained. Since pixels are missing as shown in the figure, by performing interpolation processing, a hologram with no pixel dropout is generated. From the four interpolated holograms, object light on the hologram surface can be restored using Eq. (4.35). By back-propagating this to the original position, the object light can be regenerated.

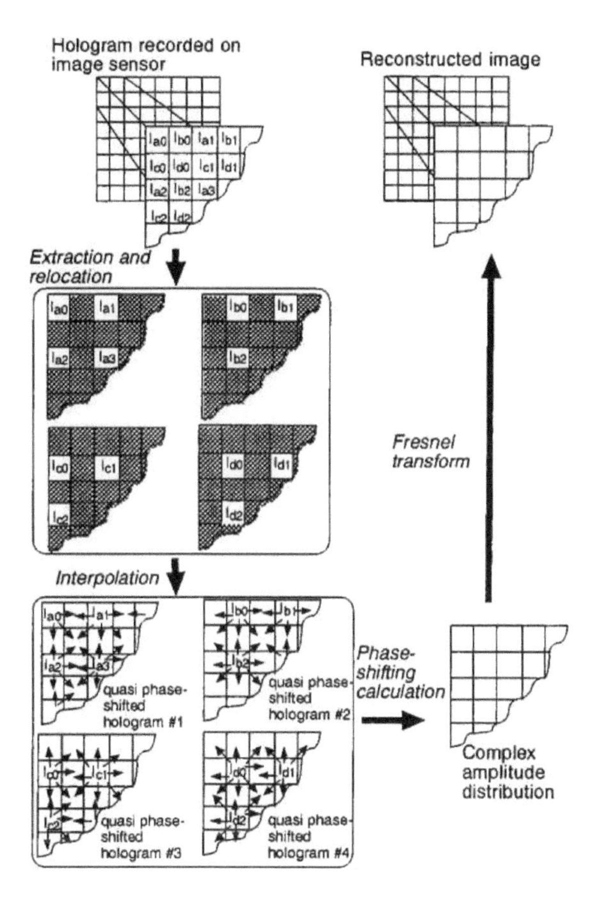

FIGURE 4.26 Reconstruction algorithm for parallel phase-shifting digital holography. Reproduced from [6] , with the permission of AIP Publishing.

By using parallel phase-shifting digital holography in combination with a high-speed camera, it is possible to perform three-dimensional imaging of high-speed phenomena [7]. Figure 4.27 shows a reconstructed image obtained by recording a jet of transparent compressed gas ejected from a spray nozzle at a recording speed of 180,000 fps. Though it cannot observe such a transparent object with the amplitude information of the object light, if phase information is used, 3D moving image imaging of a transparent subject like a gas can be performed like Figure 4.27.

Other examples of one-step phase-shifting digital holography using the Talbot effect [106, 107], random reference light [108], and inclined reference light [109] have also been proposed.

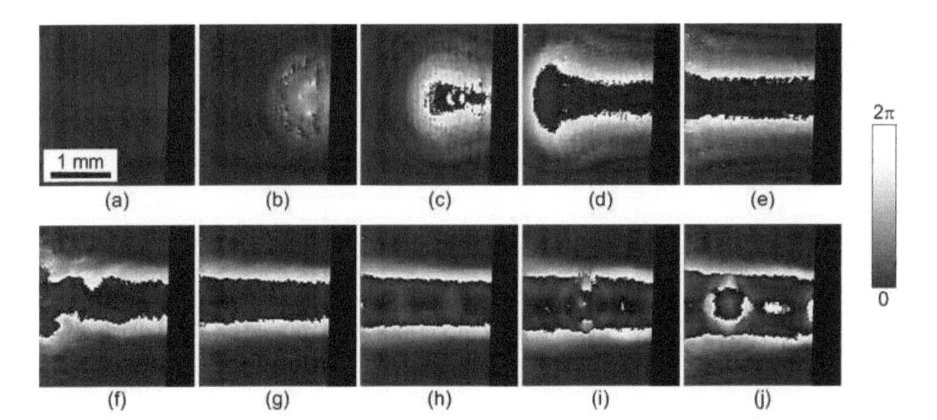

FIGURE 4.27 Reconstructed image obtained by recording a jet of transparent compressed gas ejected from a spray nozzle at a recording speed of 180,000 fps. Reprinted from Ref. [7] with permission, OSA.

4.7 AUTOFOCUSING ALGORITHM

For measuring objects in three-dimensional space by digital holography, we need to know the depth position of the recorded objects in advance. Subsequently, diffraction is calculated at that position.

If we do not know the depth position in advance, depth search is performed by multiple diffraction calculations while changing different depth parameters z. The amplitude of the diffraction calculation is expressed as

$$o_z(x, y) = |\text{Prop}_z[u_h(x_h, y_h)]| \tag{4.45}$$

where $u_h(x_h, y_h)$ is a hologram and $o_z(x, y)$ is the amplitude of the reconstructed image. We detect the most focused depth position among the results o_z by certain focus metrics. This is called **autofocusing**.

A number of focus metrics in digital holography have been proposed; e.g., entropy-based methods [110–112], a Fourier spectrum-based method [113] and a wavelet-based method [114]. Moreover, the Laplacian, variance, gradient, and Tamura metrics for the amplitude of a reconstructed image have been used as focus metrics [8, 113].

The **Laplacian metric** C_L is defined as

$$C_L = \int\int |\nabla^2 o_z(x, y)|^2 dxdy \tag{4.46}$$

where ∇^2 is the Laplacian.

The **variance metric** C_V is defined as

$$C_V = \int\int |o_z(x, y) - \bar{o_z}(x, y)| dxdy \tag{4.47}$$

where $\bar{o_z}(x, y)$ is the average of $o_z(x, y)$.

The **gradient metric** C_G is defined as

$$C_G = \int\int \sqrt{\left(\frac{\partial o_z(x, y)}{\partial x}\right)^2 + \left(\frac{\partial o_z(x, y)}{\partial y}\right)^2} dxdy. \tag{4.48}$$

The **Tamura metric** C_T is calculated as

$$C_T = \frac{\sigma_{o_z}}{\bar{o_z}}, \tag{4.49}$$

where σ_{o_z} and $\bar{o_z}$ are the standard deviation and average of the amplitude, respectively.

Almost all the metrics measure the sharpness of the reconstructed images and judge the sharpest image by the focus image. The Tamura metric is the easiest method for finding the global solution of the depth position compared to the Laplacian, variance and gradient metrics [8].

Autofocusing using these metrics is performed as follows: we calculate multiple reconstructed images o_z using the diffraction calculation by changing the depth position z from z_1 to z_2 with a step δz. Subsequently, we calculate metrics for the reconstructed images. We can find the focused depth position at the maximum metrics.

Figure 4.28 shows autofocusing results using the Laplacian, variance, gradient, and Tamura metrics [8]. The search range is $z_1 = 400$ mm to $z_2 = 1,400$ mm. Although the Laplacian, variance, and gradient metrics detect the focus position at the maximum value, these metrics have many local maxima that might hinder the correct depth search. In contrast, the Tamura metric can detect the focus position without local maxima. In addition, the detected position by the Tamura metric is more accurate than the positions detected by other methods.

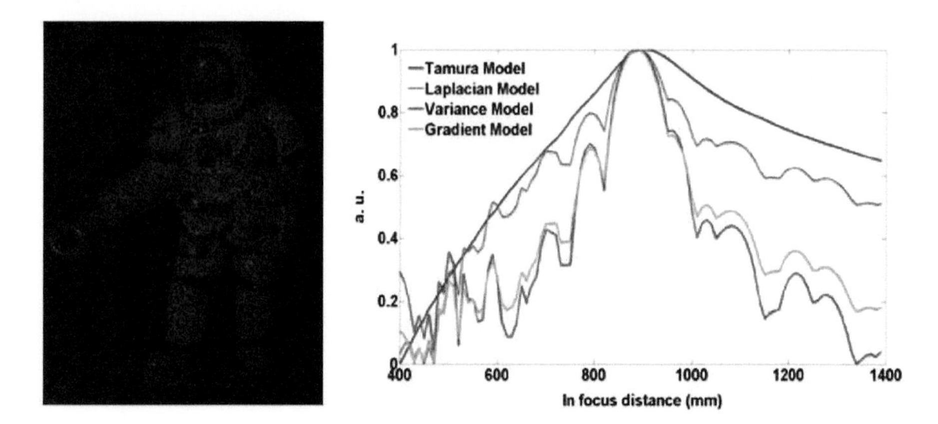

FIGURE 4.28 Autofocusing using the Laplacian, variance, gradient, and Tamura metrics. Reprinted from Ref. [8] with permission, OSA.

5 Applications of computer holography

In the previous chapters, we introduced three-dimensional display (electronic holography) and measurement (digital holography) using computer holography. Computer holography has many other applications. In this section, we introduce phase retrieval algorithms that are different from digital holography but are often used in combination with computer holography. We also introduce holographic memory and holographic projection. Finally, we introduce applications of deep learning to computer holography.

5.1 PHASE RETRIEVAL ALGORITHM

Holography records the complex amplitude of an object light in interference fringes by interfering with an object light and a reference light. Apart from holography, as shown in Figure 5.1, techniques for recovering the complex amplitude of the object light using only the diffraction patterns of the object light and known information without using interference have been studied. This technique is called the **phase retrieval algorithm** [115]. Since the phase retrieval algorithm was originally developed in the field of measurement, it is compatible with digital holography. In addition, by using the phase retrieval algorithm framework in hologram calculation, it can also be used for optimizing holograms with better image quality and holograms with higher diffraction efficiency. In this way, the phase retrieval algorithm is an important technique in computer holography. Here, we introduce several phase retrieval algorithms.

5.1.1 GERCHBERG–SAXTON ALGORITHM

The **Gerchberg–Saxton algorithm** (**GS algorithm**) developed by Gerchberg and Saxton, recovers the complex amplitude of an object light from the diffraction intensity pattern in combination with the amplitude of the object light, which is known information, detected by an image sensor. The phase information of the object light disappears in the diffraction intensity pattern used in the phase retrieval algorithm.

Various methods have been proposed for the phase retrieval algorithm, but many methods are improved based on the GS algorithm. Figure 5.2 shows the flow of the GS algorithm calculation. The GS algorithm recovers the phase in the following steps.

The known information of the original GS algorithm is the amplitude of an object light $a_o(x_o, y_o)$, the diffraction intensity pattern $I(x_d, y_d)$. (x_o, y_o) is the coordinate of the object plane, and (x_d, y_d) is the coordinate of the

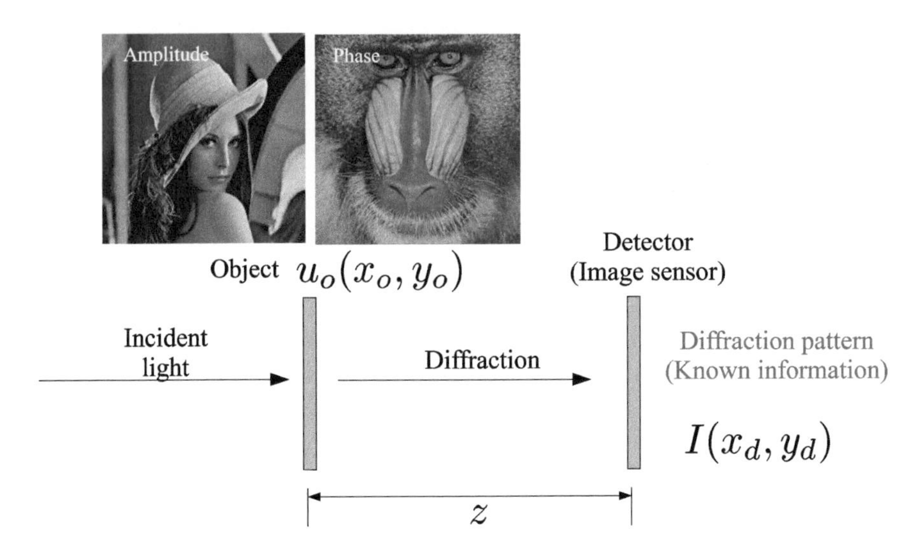

FIGURE 5.1 Phase retrieval algorithm.

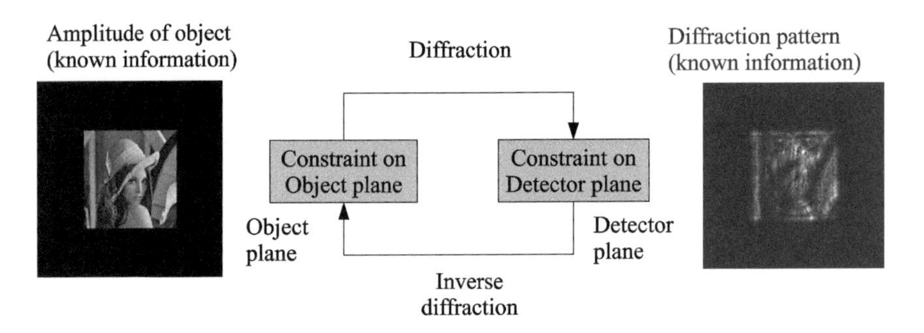

FIGURE 5.2 Calculation flow of the GS algorithm.

detector plane.[i] It is necessary to measure the known information (diffraction pattern and amplitude) in advance. According to Figure 5.2, the unknown phase information $\phi_o(x_o, y_o)$ of the object is recovered.

1. The initial complex amplitude of the object can be expressed as follows:

$$u_o(x_o, y_o) = a(x_o, y_o) \exp(i\phi_o(x_o, y_o)). \tag{5.1}$$

   Since the initial phase of the object $\phi_o(x_o, y_o)$ is unknown, it often gives a random phase distribution of 0 to 2π radians.

[i]The subscripts o and d are denote an object and detector, respectively.

2. Next, the diffraction calculation from this initial complex amplitude to the detector plane is performed by

$$u_d(x_d, y_d) = \text{Prop}_z[u_o(x_o, y_o)]. \tag{5.2}$$

3. We perform constraint in the detector plane. For $u_d(x_d, y_d)$, we maintain the phase and replace only the amplitude with the known information $I(x_d, y_d)$.

$$\begin{aligned} u'_d(x_d, y_d) &= \sqrt{I(x_d, y_d)} \exp\left(i \tan^{-1} \frac{\text{Im}\{u_d(x_d, y_d)\}}{\text{Re}\{u_d(x_d, y_d)\}}\right) & (5.3) \\ &= \sqrt{I(x_d, y_d)} \frac{u_d(x_d, y_d)}{|u_d(x_d, y_d)|}. & (5.4) \end{aligned}$$

Calculating the square root for $I(x_d, y_d)$ is required because the detected diffraction intensity can be regarded as the absolute square of the amplitude of the object light on the detector plane.

4. We compute the inverse diffraction of $u'_d(x_d, y_d)$ to find the complex amplitude at the object plane. The calculation is performed by

$$u'_o(x_o, y_o) = \text{Prop}_{-z}[u'_d(x_d, y_d)]. \tag{5.5}$$

5. We perform constraint in the object plane. We calculate new object light $u_o(x_o, y_o)$ by maintaining the phase of $u'_o(x_o, y_o)$ and replacing only the amplitude of $u'_o(x_o, y_o)$ with the known information $a(x_o, y_o)$.

$$\begin{aligned} u_o(x_o, y_o) &= a(x_o, y_o) \exp\left(i \tan^{-1} \frac{\text{Im}\{u'_o(x_o, y_o)\}}{\text{Re}\{u'_o(x_o, y_o)\}}\right) & (5.6) \\ &= a(x_o, y_o) \frac{u_d(x_d, y_d)}{|u_d(x_d, y_d)|}. & (5.7) \end{aligned}$$

By repeating steps 2 to 5, the phase $\phi_o(x_o, y_o)$ of the object light gradually recovers.

Figure 5.3 is the results showing the effect of the GS algorithm by simulation. Figure 5.3(a) was reproduced directly from the diffraction intensity pattern by inverse diffraction calculation without using the GS algorithm. Neither the amplitude nor the phase of the object can be recovered. Figures 5.3(b), (c) and (d) show the result of performing the GS algorithm once, ten times, and 100 times, respectively. As the number of iterations increases, the amplitude and phase of the object light are recovered.

The intuitive reason why the object light can be recovered by the GS algorithm is that the unknown phase component of the object light gradually coincides with the known information (the intensity information of the object light and the diffraction intensity pattern) through the iterative calculation of steps 2 to 5.

As mentioned above, the GS algorithm iterates the diffraction calculation and inverse diffraction calculation. In a short wavelength light ray (X-ray

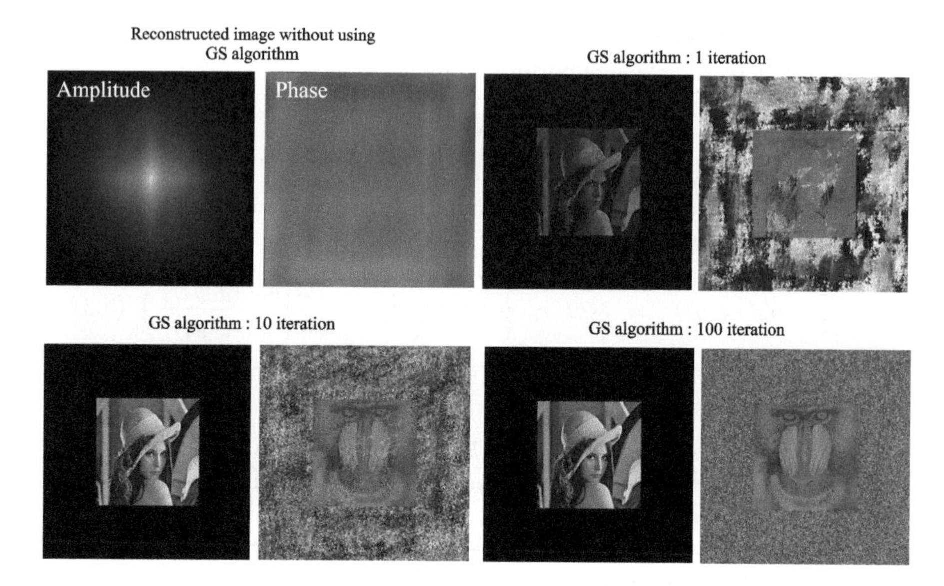

FIGURE 5.3 Reconstructed images with/without using the GS algorithm.

or electron beam), the distance between the detector and the object is longer than the wavelength, so that diffraction can be regarded as Fraunhofer diffraction,[ii] which can be calculated by the Fourier transform. Therefore, since diffraction calculation and inverse diffraction calculation can be simply replaced with the Fourier transform, the GS algorithm is also called the **Fourier iterative method**.

ERROR REDUCTION METHOD

In the original GS algorithm, the amplitude of object light is required for known information. However, there are many cases in which the amplitude of the object light is not known beforehand. The **error reduction method** [115] uses the known information below instead of the amplitude of the object light.

- Support
- Nonnegative constraint

Support refers to the area where objects exist. The **nonnegative constraint** is the known information that the amplitude of the recovered object light does not take a negative value. The error reduction method can be implemented by replacing the object plane constraint in step 5 of the GS algorithm

[ii]See Section 2.4.

described in Section 5.1.1 with either support or a non-negative constraint (or both).

Figure 5.4 shows the calculation flow of the error reduction method and the simulation result of the error reduction method. There is noise in the recovered amplitude and phase, but both the amplitude and phase can be observed.

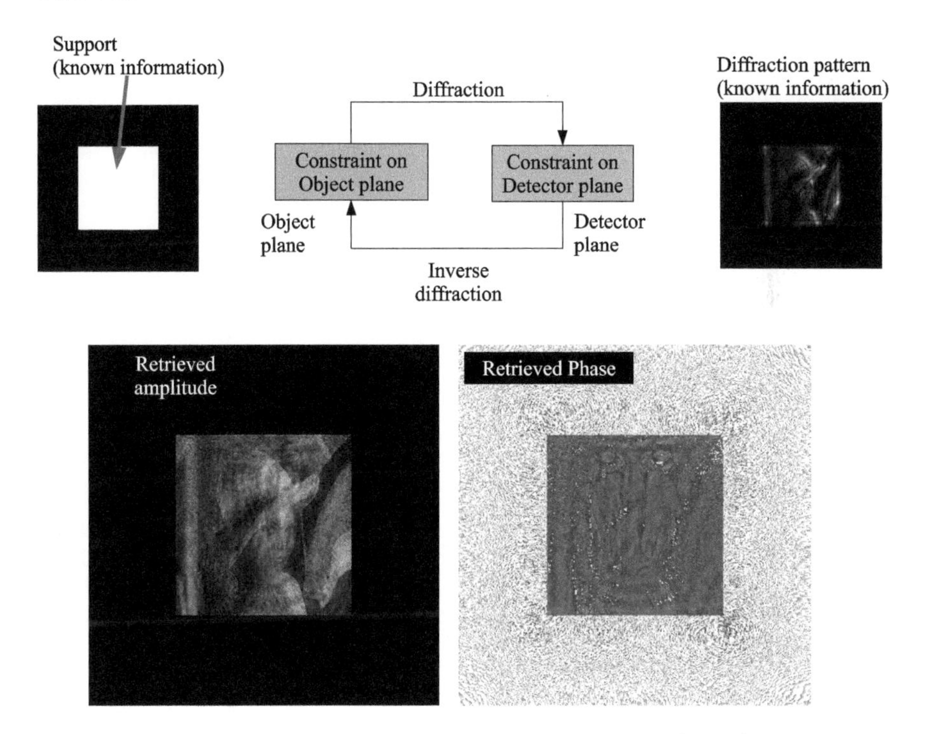

FIGURE 5.4 Calculation flow of the error reduction method (above) and the simulation results of the error reduction method (below).

HYBRID INPUT-OUTPUT ALGORITHM

The **hybrid input–output algorithm** (**HIO algorithm**) [115] is a family of the GS algorithm in which the object plane constraint in step 5 of the GS algorithm is replaced by the following update:

$$u_o(x_o, y_o) = \begin{cases} u'_o(x_o, y_o) & \text{(within support)} \\ u_o(x_o, y_o) - \beta u'_o(x_o, y_o) & \text{(outside of support)} \end{cases}$$

β is a weight that takes a range between 0 and 1, and is often set to about 0.9 empirically. In the error reduction method, we set the non-support area to zero, whereas the HIO method avoids stagnation in phase recovery, which often becomes a problem in iterative calculations, by taking the error of $u_o(x_o, y_o)$ and $u'_o(x_o, y_o)$.

5.1.2 PHASE RETRIEVAL ALGORITHM USING MULTIPLE DIFFRACTION PATTERNS

A phase retrieval algorithm using multiple diffraction patterns has been introduced [9]. This method records multiple diffraction intensity patterns from the object plane while gradually shifting an imaging sensor in the axial direction, as shown in Figure 5.5. Then, using these diffraction patterns as known information, we recover the complex amplitude of the object.

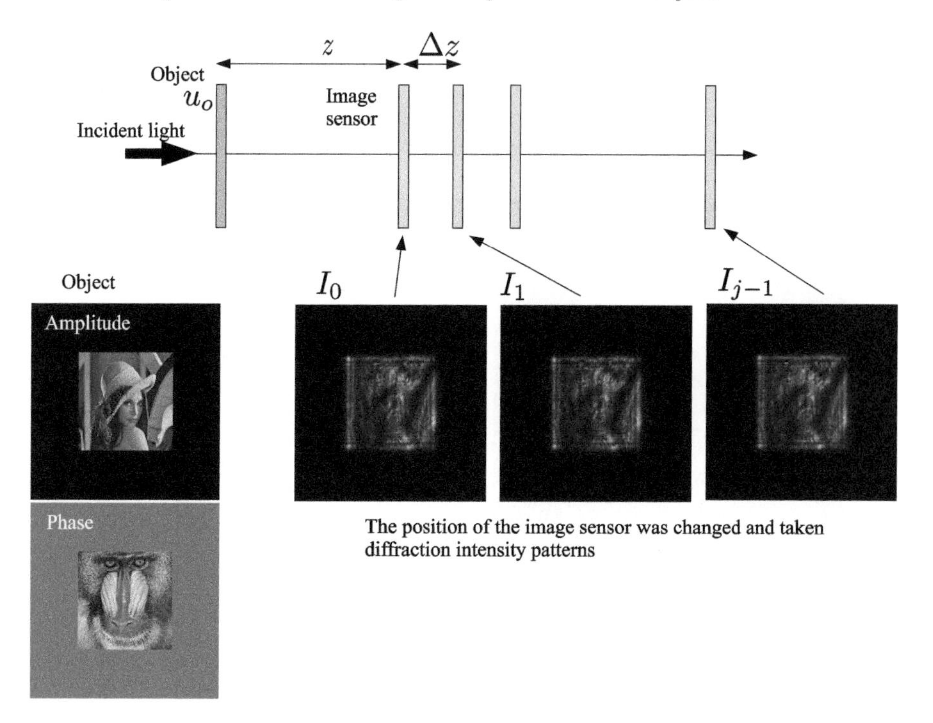

FIGURE 5.5 Phase recovery algorithm using multiple diffraction patterns.

We record j diffraction patterns while shifting the image sensor by Δ_z in the axial direction. The diffraction patterns have different patterns. The following iterative calculation is performed using these patterns as known information to restore the complex amplitude of the object u_o.

1. A complex amplitude of the object light at a distance of z away from the object can be written as

$$u_0(x, y) = \sqrt{I_0(x, y)} \exp(i\phi_0(x, y)) \tag{5.8}$$

using the diffraction pattern $I_0(x, y)$ captured by the image sensor at the initial position. $\phi_0(x, y)$ is the initial phase of the object light at this position and it is set to 0 [9].

2. When $u_0(x, y)$ is propagated at Δ_z, we write the complex amplitude as

$$u_1(x, y) = \mathrm{Prop}_{\Delta_z}[u_0(x, y)]. \tag{5.9}$$

Only the amplitude of $u_1(x, y)$ is replaced by the known information of the diffraction pattern $I_1(x, y)$ at this position (the position $z+\Delta_z$ from the object), which has been measured beforehand. The complex amplitude of the object light is written by

$$\begin{aligned} u_1'(x, y) &= \sqrt{I_1(x, y)} \exp\left(i \tan^{-1} \frac{\mathrm{Im}\{u_1(x, y)\}}{\mathrm{Re}\{u_1(x, y)\}}\right) & (5.10) \\ &= \sqrt{I_1(x, y)} \frac{u_1(x, y)}{|u_1(x, y)|}. & (5.11) \end{aligned}$$

3. If we perform the same calculation as in the previous step, the complex amplitude of object light at position $z + j\Delta_z$ is written by

$$\begin{aligned} u_j'(x, y) &= \sqrt{I_j(x, y)} \exp\left(i \tan^{-1} \frac{\mathrm{Im}\{u_j(x, y)\}}{\mathrm{Re}\{u_j(x, y)\}}\right) & (5.12) \\ &= \sqrt{I_j(x, y)} \frac{u_j(x, y)}{|u_j(x, y)|}, & (5.13) \end{aligned}$$

where $u_j(x, y)$ is defined by

$$u_j(x, y) = \mathrm{Prop}_{\Delta_z}[u_{j-1}(x, y)]. \tag{5.14}$$

4. The object light $u_o(x, y)$ we want to obtain exists at the position of $-(z + j\Delta_z)$ from the image sensor. We can find $u_o(x, y)$ by back-propagating $u_j'(x, y)$ as

$$u_o(x, y) = \mathrm{Prop}_{-(z+j\Delta_z)}[u_j(x, y)]. \tag{5.15}$$

Figure 5.6 is the simulation result when the number of the measured diffraction intensity patterns is 8. Although some noise is seen in the reconstructed image, it can be seen that both the amplitude and phase of the object beam can be restored. The image quality of the reconstructed image can be further improved by increasing the number of known diffraction intensity patterns.

Figure 5.7 is the experimental result in which 30 diffraction intensity patterns are used. It can be seen that the amplitude and phase of the object light can be restored from the diffraction intensity patterns.

5.2 HOLOGRAPHIC MEMORY

Compact disks (CDs), digital versatile disks (DVDs) and Blu-Ray discs are widely used as optical memory. They record digital data as minute pits on

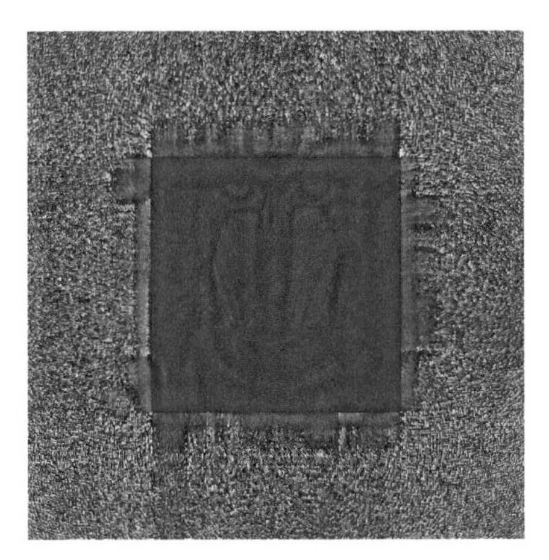

FIGURE 5.6 Simulation result of the phase retrieval algorithm using multiple diffraction patterns when the number of the measured diffraction intensity patterns is 8.

a plastic disc. The laser light focused by lenses is irradiated to the pit, and the digital data is read by detecting the reflected light. In order to increase the storage capacity, the pits must be miniaturized. Accordingly, a small laser spot must be generated.

The size of the laser spot d is decided by

$$d \propto \frac{\lambda}{\mathrm{NA}} \tag{5.16}$$

where λ is the wavelength of the laser and NA is the **numerical aperture** of the lens. For Blu-Ray discs, the minimum pit is 0.15 μm, and to achieve higher densities, it is necessary to use a laser with a shorter wavelength and a lens with a higher NA. However, this has reached its limit.

Holographic memory converts digital data to be recorded into a two-dimensional pattern (**data page**) and records it in a recording medium as a hologram. A holographic memory is generally a volumetric optical memory which records interference fringes in both the lateral and thickness direction of the recording medium. The main features of holographic memory are summarized below.

- Since digital data can be written and read as two-dimensional patterns, the access speed is fast.
- A plurality of data pages can be recorded in the same recording area by the multiplex recording characteristic of the hologram, so holographic memory has a large storage capacity.
- Since holograms can be redundantly recorded, even if the recording medium is slightly missing, it is expected to restore the recorded data.

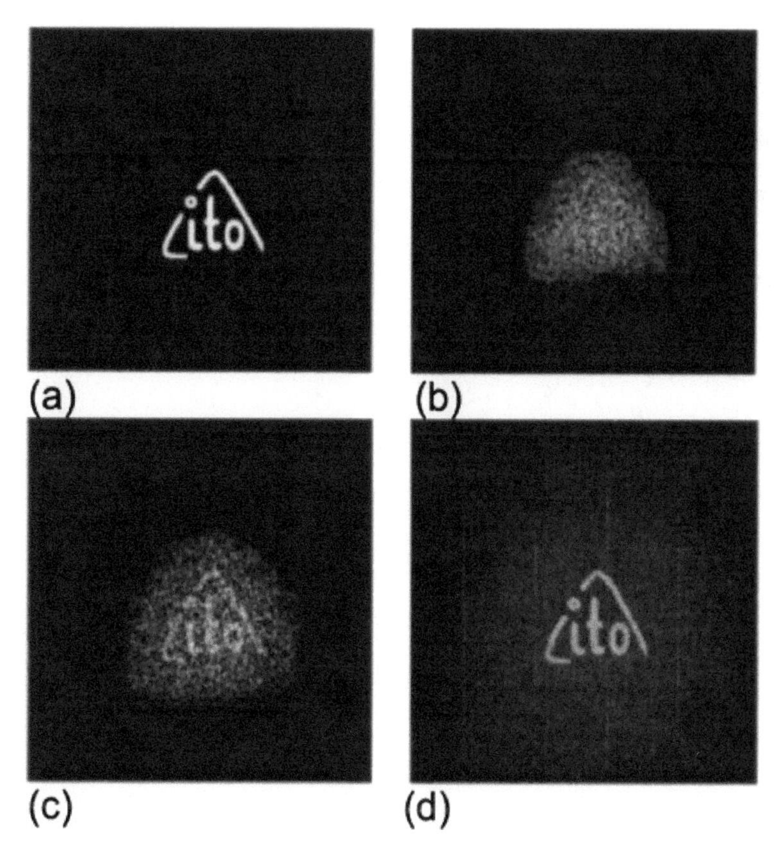

FIGURE 5.7 Experimental results of the phase retrieval algorithm using multiple diffraction patterns in which 30 diffraction intensity patterns are used. Reprinted from Ref. [9] with permission, OSA.

The outline of holographic memory is shown in Figure 5.8. The digital data to be recorded is rearranged as data page, which is displayed on a spatial light modulator such as an LCD. The spatial light modulator is irradiated with laser light to obtain object light (also called **signal light** in the field of holographic memory). In order to increase the recording density, the object light is reduced by lenses and recorded in a small area of the recording medium. By interfering with the object light and the reference light on the recording medium, it is possible to record the data page as a hologram. At this time, when the recording medium has a sufficient thickness as compared with the wavelength, the interference fringes exist in both of the lateral and depth direction and become a volume hologram.

Representative holographic memory recording materials are as follows.

- Photosensitive material

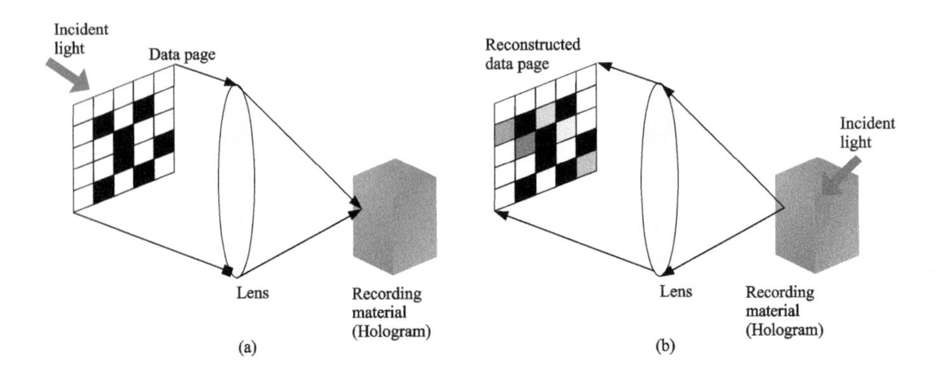

FIGURE 5.8 Outline of holographic memory.

- Photorefractive material
- Photopolymer

A **photosensitive material** is a recording medium often used in analog holography and was used in early studies of holographic memory. A **photorefractive material** is a material whose refractive index changes according to the intensity of light incident on this material. It is rewritable. A **photopolymer** is also often used. The photopolymer cures at specific wavelengths. It can only be written once.

In order to read the recorded data, the hologram is irradiated with reproduction light, and then the reconstructed data page is captured by an image sensor. The history of holographic memory is old. Its origin dates back to Gabor's 1949 paper. Here, the outline of holographic memory will be described.

5.2.1 VOLUME HOLOGRAM

In holographic memories, it is customary to record data page as a **volume hologram** (a.k.a., **thick hologram**). Unlike a thin hologram, in a volume hologram, object light can only be reconstructed at the same wavelength and incident angle as at the time of recording. The phenomenon that diffracted light appears only under such specific conditions is called **Bragg diffraction**.

As shown in Figure 5.9(a), consider the interference fringes in the hologram when the object light u_o (wave vector $\mathbf{k_o}$) and the reference light u_r (wave vector $\mathbf{k_r}$) are incident in the recording medium with the thickness t. Here, for the sake of simplicity, assuming that the object beam and the reference beam have plane waves with the amplitude of 1, we can write them as

$$\begin{aligned} u_o &= \exp(i\mathbf{k_o} \cdot \mathbf{r}) = \exp(ik(x \sin\theta_o + z\cos\theta_o)) \\ u_r &= \exp(i\mathbf{k_r} \cdot \mathbf{r}) = \exp(ik(-x \sin\theta_r + z\cos\theta_r)) \end{aligned} \tag{5.17}$$

where $\mathbf{r} = (x, z)$ is the position vector and $k = 2\pi/\lambda$ is the wave number (λ is the wavelength).

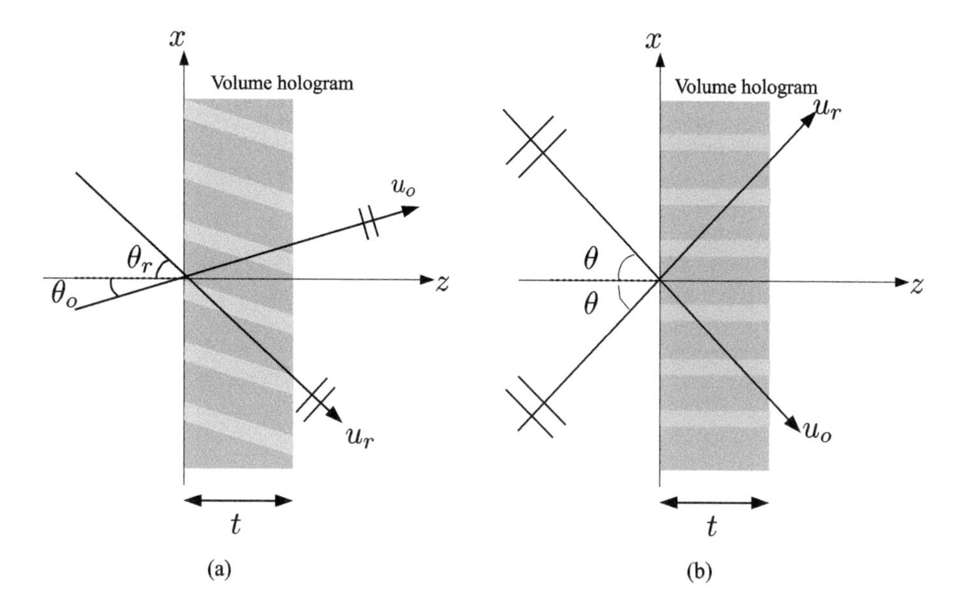

FIGURE 5.9 Volume hologram recording.

Therefore, the hologram is expressed as

$$\begin{aligned} I(x,z) &= |u_o + u_r|^2 = 2 + 2\cos((\mathbf{k_o} - \mathbf{k_r}) \cdot \mathbf{r}) \\ &= 2 + \underbrace{2\cos(k(x(\sin\theta_o + \sin\theta_r) + z(\cos\theta_o - \cos\theta_r)))}_{\text{Interference fringes formed in the recording medium}}. \end{aligned} \tag{5.18}$$

The second term indicates the interference fringe. The condition of constructive interference is when the phase of the second term is $2\pi n$ (n is an integer), so that we can write the condition as

$$k(x(\sin\theta_o + \sin\theta_r) + z(\cos\theta_o - \cos\theta_r)) = 2\pi n. \tag{5.19}$$

The interference fringes satisfying this equation are recorded in the x and z axis directions of the recording medium.

When the incident angle of the object light and reference light are the same as shown in Figure 5.9(b) ($\theta_o = \theta_r = \theta$), the term relating to z of Eq. (5.18) becomes zero. The interference fringes of the hologram are expressed by

$$x = \frac{n\pi}{k\sin\theta} = \frac{n\lambda}{2\sin\theta}. \tag{5.20}$$

It changes in the x-axis direction and has a uniform structure in the z-axis direction. The interference period d is expressed by

$$d = \frac{\lambda}{2\sin\theta}. \tag{5.21}$$

Let us intuitively consider the reconstruction from the hologram of Figure 5.9(b) [13]. When a plane wave is incident to this hologram at an angle α, scattered lights from each layer of the hologram are generated as shown in Figure 5.10(a), and the recorded object light is reconstructed as shown in Figure 5.10(a). The direction in which the object lights are strengthened is considered when the following conditions are met.

- Object lights from different positions of a layer is constructive each other, as shown in Figure 5.10(b)
- Object lights from different layers are constructive, as shown in Figure 5.10(c)

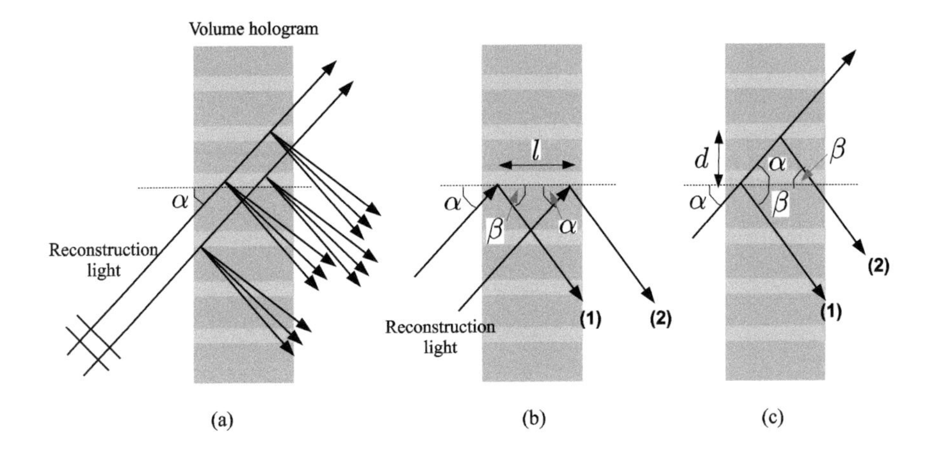

FIGURE 5.10 Intuitively understanding of the volume hologram reconstruction.

In Figure 5.10(b), we obtain the conditions for making constructive object light. Assuming that the distance of the reconstruction light irradiated to different positions is l, the optical path of (2) ($l \cos \alpha$) is longer than that of (1) for the incidence. For the reflection, the optical path of (2) is shorter than that of (1) ($l \cos \beta$). Therefore, the condition for the constructive object light is written as

$$l(\cos \alpha - \cos \beta) = m\lambda \tag{5.22}$$

where m is an integer.

In addition, in Figure 5.10(c), the optical path of (2) ($d \sin \alpha$) is longer than that of (1) for incidence. For the reflection, the optical path of (2) is shorter than that of (1) ($d \sin \beta$). Therefore, the condition for the constructive object light is written as

$$d(\sin \alpha + \sin \beta) = m\lambda. \tag{5.23}$$

The constructive condition under which Eq. (5.22) holds for arbitrary l is only

$$\alpha = \beta. \tag{5.24}$$

Substituting this for Eq. (5.23) results in

$$2d \sin \alpha = \lambda. \tag{5.25}$$

Therefore, we obtain

$$\alpha = \beta = \sin^{-1} \frac{\lambda}{2d}. \tag{5.26}$$

Substituting Eq. (5.21) for Eq. (5.26) results in

$$\alpha = \beta = \sin^{-1} \frac{\lambda}{2\frac{\lambda}{2\sin\theta}} = \theta. \tag{5.27}$$

This equation shows that the object light can be reconstructed when the same reference light as the hologram recording is incident to the hologram. This angle is called **Bragg angle**. The reconstructed object light has the highest **diffraction efficiency** at this angle.[iii] When the angle deviates from this angle, the object light rapidly decreases.

5.2.2 REPRESENTATION OF A VOLUME HOLOGRAM BY WAVEVECTOR

A volume hologram can be expressed by using the wavevectors of the reconstruction light and the reconstructed object light, and the direction of the grating (**grating vector**). In Section 5.2.1, the volume hologram was recorded by using the same angles θ for the object light and the reference light.

Consider the reconstruction from a volume hologram when object light and reference light are incident at different angles as shown in Figure 5.9(a). As shown in Figure 5.11, when the reconstruction light with the wavevector $\mathbf{k}_r$ is irradiated to the volume hologram with the grating vector $\mathbf{d}$, the reconstructed object light with the wavevector $\mathbf{k}_o$ is reproduced. The wavevector $\mathbf{k}_o$ is expressed as

$$\mathbf{k}_o = \mathbf{k}_r + \mathbf{d}. \tag{5.28}$$

The direction of the reconstructed object light is determined by a direction in which the three vectors are in contact with the circle (**Ewald sphere**) shown in Figure 5.11(b), and at this time it has high diffraction efficiency.

When the volume hologram is irradiated with reconstruction light $\mathbf{k'}_r$ that is different from the reference light at the time of recording, the diffraction efficiency of the reconstructed object light rapidly decreases. As shown in Figure 5.12(a), the wavevector expression does not fit in the Ewald sphere but is shifted by $\Delta\mathbf{k}$. This can be expressed by using wave vectors and a grating vector as

$$\mathbf{k}_o = \mathbf{k'}_r + \mathbf{d} - \Delta\mathbf{k}. \tag{5.29}$$

[iii] It indicates how much of the energy of incident light is used as diffracted light. When the diffraction efficiency is 1 (100 %), the energy of all incident light becomes diffracted light.

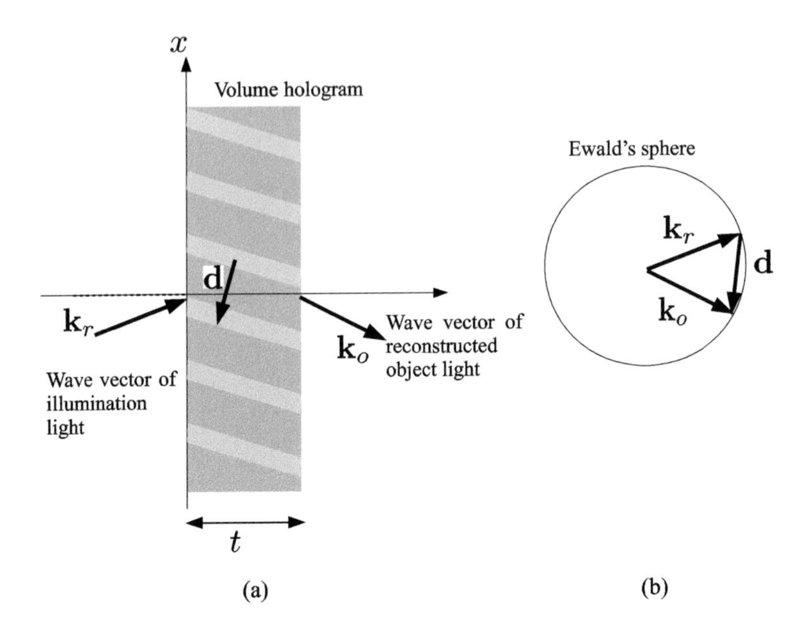

FIGURE 5.11 Direction of the reconstructed object light is determined by a direction in which the three vectors are in contact with the Ewald sphere.

The diffraction efficiency η is expressed by [116, 117][iv]

$$\eta \propto \text{sinc}^2\left(\frac{t\Delta\mathbf{k}\cdot\mathbf{e}_z}{2}\right) \tag{5.30}$$

where t represents the thickness of the volume hologram, $\mathbf{e}_z$ is a unit vector in the z-axis direction, and $\text{sinc}(x) = \sin(\pi x)/\pi x$ is called the **sinc function**.

As shown in Figure 5.12(a), when $\mathbf{k}'_r$ is equal to the wavevector of the reference light at the recording, $\Delta\mathbf{k}$ is zero. Therefore, the diffraction efficiency is maximized as shown in Figure 5.12(b). When the condition of the reconstruction light differs from the reference light at the recording, the diffraction efficiency rapidly decreases.

5.2.3 ANGULAR MULTIPLEX RECORDING METHOD

Holographic memory enables high-density recording by using a multiplex recording of multiple data pages at the same area in the recording medium. Such a recording method is called **multiplex recording**, and several methods have been proposed.

- Angle multiplex recording method

[iv]It can be derived from coupled-wave theory by Kogelnik (Coupled-Wave Theory), but it exceeds the scope of this book. Here, only the result is shown.

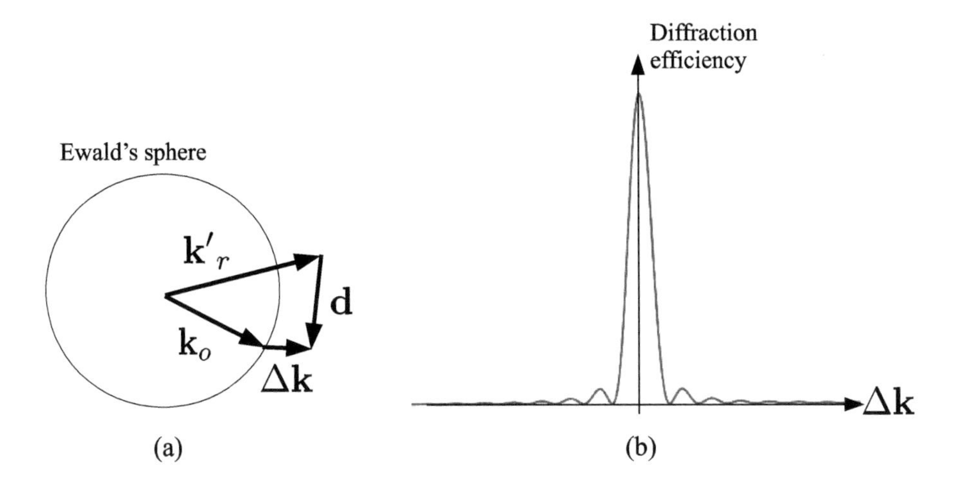

FIGURE 5.12 Volume hologram is irradiated with the reconstruction light $\mathbf{k}'_r$ that is different from the reference light at the time of recording.

- Wavelength multiplex recording method
- Shift multiplex recording method

Here we introduce the **angle multiplexing method**. A system of the angle multiplexing method is shown in Figure 5.13. A data page is displayed on a spatial light modulator, and the data page reduced by the lens is irradiated to the recording medium. By interfering the object light with the reference light, it is possible to record the data page in the recording medium. To read the recorded data page, the object light is reconstructed by irradiating the reconstruction light at the same angle as the reference light. Then, the data page can be read by capturing the reconstructed object light with an image sensor.

When other data page are recorded in the same area without changing the angles of the reference light, all the recorded data pages are simultaneously reconstructed, and crosstalk occurs.

In the angular multiplex recording method, a characteristic that the diffraction efficiency of the reconstructed object light decreases rapidly when the volume hologram deviates from the Bragg angle, is used (Figure 5.14). In Figure 5.13, when data page 1 is recorded, the reference light is set to the angle of θ_1. In the same way, when data pages 2 and 3 are recorded, the reference lights are set to the angles of θ_2 and θ_3, respectively. At this time, as shown in Figure 5.14, each angle of the reference lights is set to the angles where the diffraction efficiency of Figure 5.14 is the most degraded. When we want to reconstruct data page 1, data page 1 is reconstructed with high diffraction efficiency by irradiating the reconstruction light with the angle of θ_1, whereas

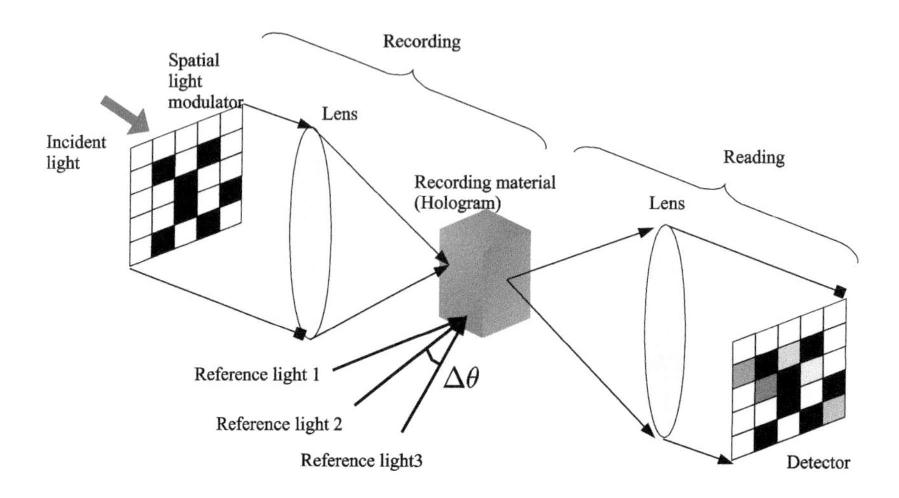

FIGURE 5.13 Angle multiplex recording method.

other data pages are not reconstructed.

By decreasing the angle difference $\Delta\theta$ between the reference lights, it can be seen that more information can be recorded. The diffraction efficiency of the angle multiplexing method [117] is written as

$$\eta \propto \mathrm{sinc}^2(\frac{\pi t \sin(\theta_r - \theta_o)\Delta\theta}{\lambda \cos\theta_o}), \tag{5.31}$$

where θ_r is the incident angle of the reference beam, and θ_o is the incident angle of the object beam (data page).

The sinc function $\mathrm{sinc}(x) = \sin(x)/x$ becomes 0 when $x = \pi$. Therefore, the diffraction efficiency η becomes smallest when the argument of the sinc function of Eq. (5.31) is π, that is,

$$\frac{\pi t \sin(\theta_r - \theta_o)\Delta\theta}{\lambda \cos\theta_o} = \pi. \tag{5.32}$$

Namely, $\Delta\theta$ is calculated by

$$\Delta\theta = \frac{\lambda \cos\theta_o}{t \sin(\theta_r - \theta_o)}. \tag{5.33}$$

Let us consider that an object light is incident on a recording medium with $t = 1$ cm at the incident angle $\theta_o = 0°$ and a reference light (wavelength 500 nm) at the right angle ($\theta_r = 90°$). In this condition, $\Delta\theta$ is 50μ radians. If the angle range of the reference light can be changed up to $30°(= \frac{30}{180}\pi \approx 0.5$ radians), 10,000 data pages can be approximately recorded in the recording medium.

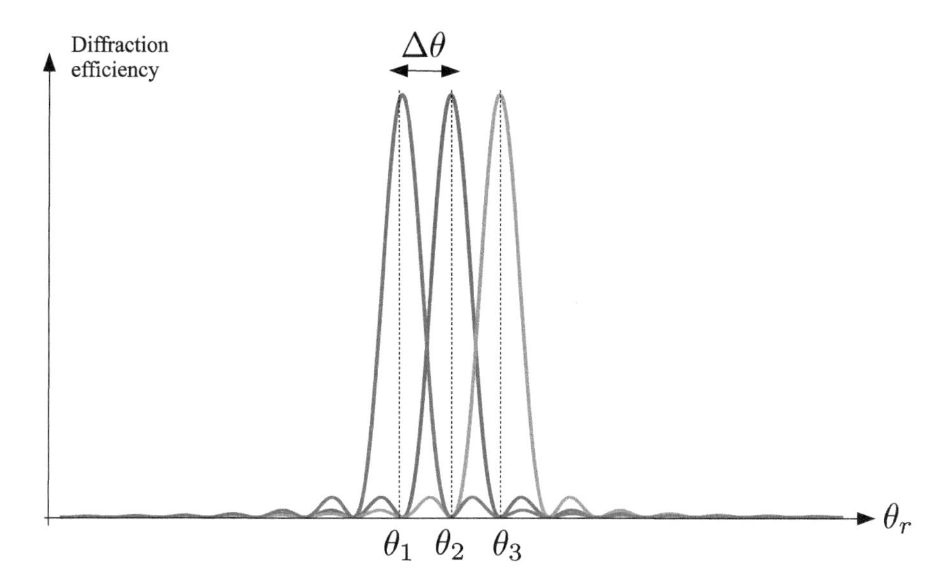

FIGURE 5.14 Angles of reference light in the angle multiplex recording method.

5.3 HOLOGRAPHIC PROJECTION

With the development of semiconductor lasers, LED light sources, display elements such as LCD panels and **micro electro mechanical systems (MEMS)**, projectors aiming at low power consumption and miniaturization are actively being developed. Recently, small projectors called micro projectors and pico projectors have been commercialized, and modular type products that can be mounted on mobile devices are also being developed. Various schemes for small projectors have been proposed, but representative ones are a method of projecting an image of an LCD panel with a lens and a laser scanning method using MEMS [118].

In particular, the latter has a feature (focus-free) that allows images to be projected to any position without a lens. Various methods other than these have been proposed for the projection technology, but **holographic projection** [119, 120] is a projector that utilizes the characteristic that the wavefront can be freely controlled.

In the holographic projection, since the hologram itself functions as a lens, it can basically be constructed without lenses, and since it does not use a lens, it is a non-aberration and suitable for miniaturization. Also, as a general laser projector, it can be expected that the projection image can be made high contrast because it uses a laser. As functions unique to holographic projection, there is multi-projection (simultaneous projection of plural images) and projection to the arbitrary curved surface. On the other hand, the disadvantages are speckle noise, zooming method, and hologram calculation time. In this section, we describe the outline of holographic projection.

5.3.1 OUTLINE OF HOLOGRAPHIC PROJECTION

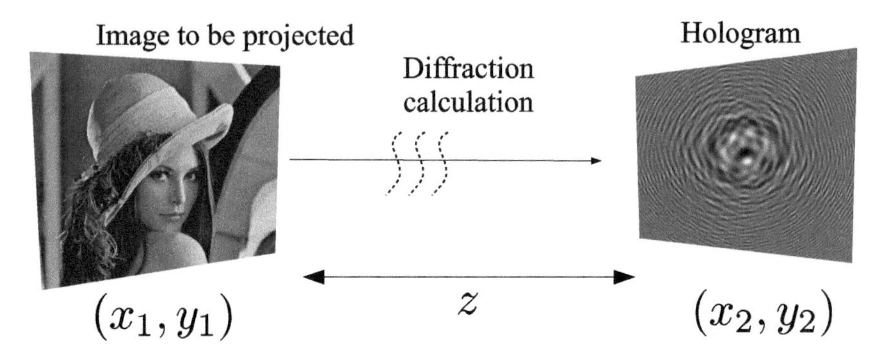

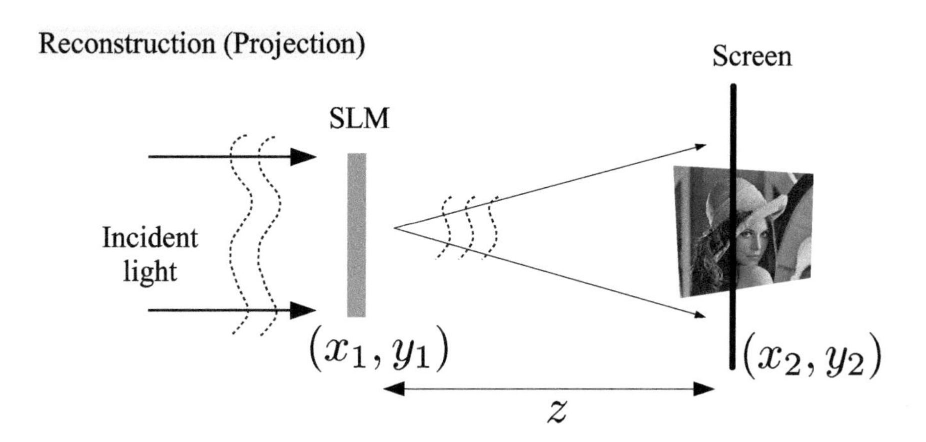

FIGURE 5.15 Outline of holographic projection.

The outline of holographic projection is shown in Figure 5.15. The holographic projection obtains a projected image in two steps: generation of a hologram and reproduction from the hologram. As shown in Figure 5.15, first, we prepare an image. The light distribution emitted from the image on the hologram is obtained as follows

$$u_2(x_2, y_2) = \text{Prop}_z[u_1(x_1, y_1)], \tag{5.34}$$

where $u_1(x_1, y_1)$ is the original image and $u_2(x_2, y_2)$ is the complex amplitude in the hologram. For this calculation, we can use the diffraction calculation introduced in Chapter 2.

Since the resultant is complex amplitude, how to display the complex amplitude on a display device is a problem. As a general display device, an amplitude modulation type capable of controlling only the amplitude of the complex amplitude or a phase modulation type capable of controlling only the phase of the complex amplitude is mainstream. Therefore, it is necessary to convert the amplitude hologram that is extracted the real part of the calculation result obtained by Eq. (5.34), or the kinoform that is the extracted the phase part of the calculation result. For example, the **amplitude hologram** $I(x_2, y_2)$ is calculated by

$$I_2(x_2, y_2) = \mathrm{Re}\{u_2(x_2, y_2)\}. \tag{5.35}$$

The **kinoform** is calculated by

$$I_2(x_2, y_2) = \tan^{-1} \frac{\mathrm{Im}\{u_2(x_2, y_2)\}}{\mathrm{Re}\{u_2(x_2, y_2)\}}. \tag{5.36}$$

Development of a display device capable of displaying complex amplitude $u_2(x_2, y_2)$ has been advanced.

Next, consider reproducing the projected image from the hologram (Figure 5.15). If a hologram is displayed on the display device and the reference light is irradiated to the device, the projection image is formed at the position of the distance z at the time of calculation. The image can be projected without the lens in the optical system.

Figure 5.16 shows projection images (computer simulation results) from complex holograms, amplitude holograms, and kinoforms when using the optical system of Figure 5.15. Since the complex hologram records the object light as it is, the reconstructed image can be almost completely reproduced, but since the amplitude hologram and kinoform recorded a part of the object beam, the projected image deteriorates.

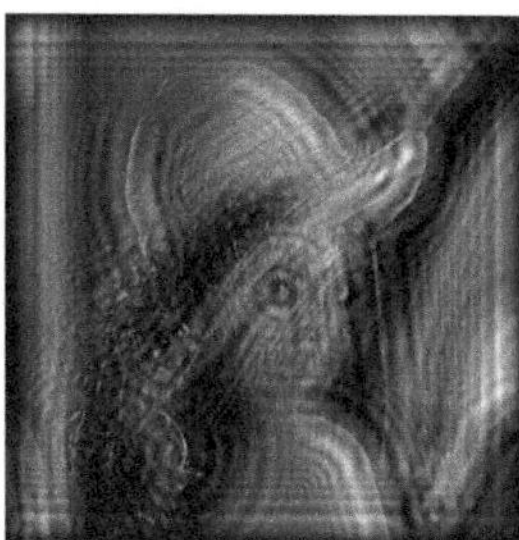

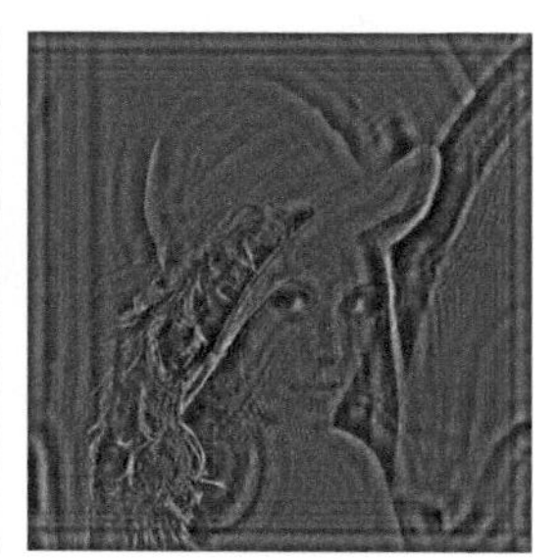

FIGURE 5.16 Projection images (computer simulation results) from complex holograms, amplitude holograms, and kinoforms when using the optical system of Figure 5.15.

This problem can be relaxed by multiplying the original image by a **random phase** distribution $\exp(i2\pi n(x_1, y_1))$ as shown in the following formula.

$n(x_1, y_1)$ is a random number ranging from 0 to 1.

$$u_2(x_2, y_2) = \mathrm{Prop}_z[u_1(x_1, y_1) \exp(i2\pi n(x_1, y_1))] \tag{5.37}$$

Random phase has the same effect as physically placing the diffuser in the immediate vicinity of the original image.

A specific effect of random phase in holography is to record the information of the original image uniformly in the hologram. For example, kinoform is a method of recording only the phase of object light, assuming the amplitude of the object light on the hologram to be constant. However, since the amplitude of the original image in the hologram greatly fluctuates, this assumption usually does not hold. Therefore, the random phase is multiplied to diffuse the amplitude information and uniformly distribute it on the hologram. If the amplitude information becomes uniform, it can be eliminated as a constant, and three-dimensional information can be reproduced only by the phase information. It is known that kinoform can be regarded as a complete hologram if each pixel of the original image has independent and random phases [13].

Figures 5.17(a) and (b) show, respectively, reconstructed images from the amplitude hologram and kinoform calculated with the random phase, which shows that the image quality is improved compared to Figure 5.16. However, the multiplication by the random phase causes **speckle noise** to appear in the projected image. Even with a general laser projector, speckle noise occurs due to irregular reflection of the laser on the rough surface of the screen. In holographic projection, in addition to this speckle generation factor, speckle due to random phase also occurs. A method of reducing this problem will be described later.

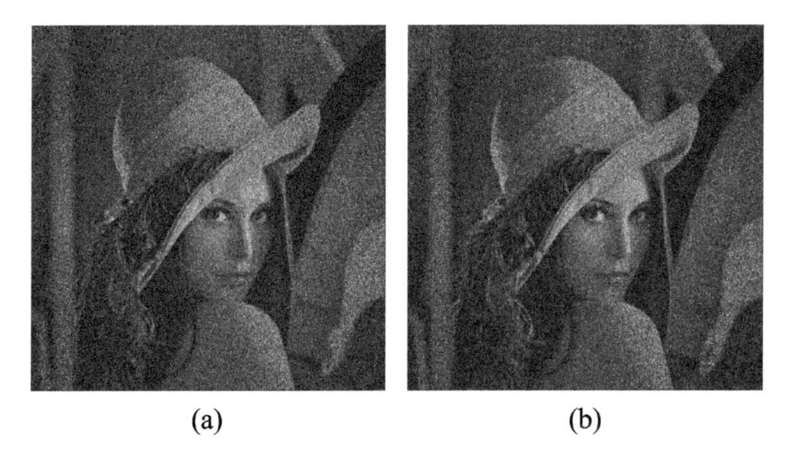

FIGURE 5.17 Reconstructed image with the random phase, (a) Amplitude hologram (b) kinoform.

5.3.2 LENSLESS ZOOMABLE HOLOGRAPHIC PROJECTOR

As shown in Figure 5.15, a holographic projector can obtain projected images without a lens in principle. Here we will introduce the holographic projection that allows zooming without using a lens [10].

In a holographic projection, the zoom function can also be realized by calculation without using a lens. There are two ways to conduct this. In Figure 5.18, the first method can easily zoom by calculating the hologram after enlarging the original image by image processing. However, the computational burden and the memory amount increase in proportion to the original image size. When we want to zoom in m times, the calculation time and memory amount increase in proportion to m^2 times. For example, when zooming 10 times, it takes 100 times more computation time and memory amount than normal holographic projection, so this method is not realistic.

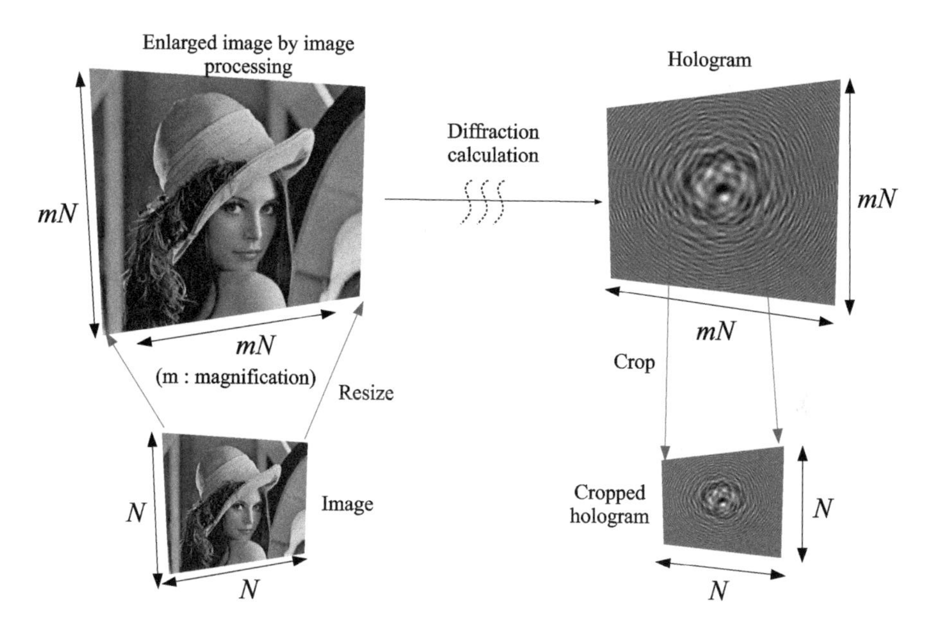

FIGURE 5.18 Lensless zoom by enlarging the size of the original images.

Figure 5.19 shows a method in which only the sampling interval of an original image is changed without increasing the number of pixels of the original image. In Figure 5.19, when the sampling interval of the hologram is set to p, the sampling interval of the original image is set to mp. This method can realize the zoom function without increasing the calculation burden and the memory amount. However, diffraction calculation, Fresnel diffraction, and the angular spectrum method described in Chapter 2, have a limitation that the sampling interval of the original image must be the same as the sampling interval of the hologram.

In recent years, several methods called **scale diffraction calculation** that can perform high-speed calculation at different sampling intervals have been proposed. Even if the sampling interval of the original image and the hologram are different by using scale diffraction calculation, high-speed calculation using FFT can be performed. The details of this calculation are described in Section 2.8.1.

Figure 5.20 shows the reconstructed image from the hologram generated using the scale diffraction calculation [10]. A phase-modulated LCD panel ($1,920 \times 1,080$ pixels) was used as the display device of the hologram. Since the pixel interval and the size of this display are 8 μm and 15 mm $\times$ 9 mm, respectively, the sampling interval of the hologram is fixed to $p = 8$ μm and the sampling interval of the original image is changed from 2 μm to 18 μm (the magnification of the projected image at this time is 0.25 to 2.25 times). We can see that zooming can be performed without using a lens from the projected image.

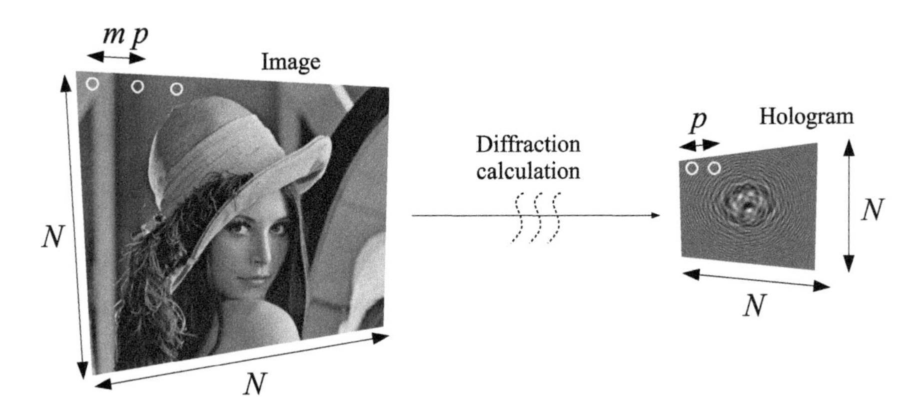

FIGURE 5.19 Lensless zoomable method by changing sampling pitches

5.3.3 SPECKLE SUPPRESSION

A screen used for projection has a rough surface with irregularities in terms of wavelength size. When a laser beam is irradiated to the screen, the light reflected by the irregularities interferes randomly and an irregular bright spot appears. This is called **speckle noise**.

In holographic projection, speckle noise due to random phase occurs in addition to speckle noise generated in the rough surface of the screen. In this section, we describe a method for reducing speckle noise.

In holographic projection, the **Gerchberg–Saxton algorithm** (**GS algorithm**) is often used to reduce this speckle noise. Although the GS algorithm was originally developed as a technique (see Section 5.1.1) to restore the phase

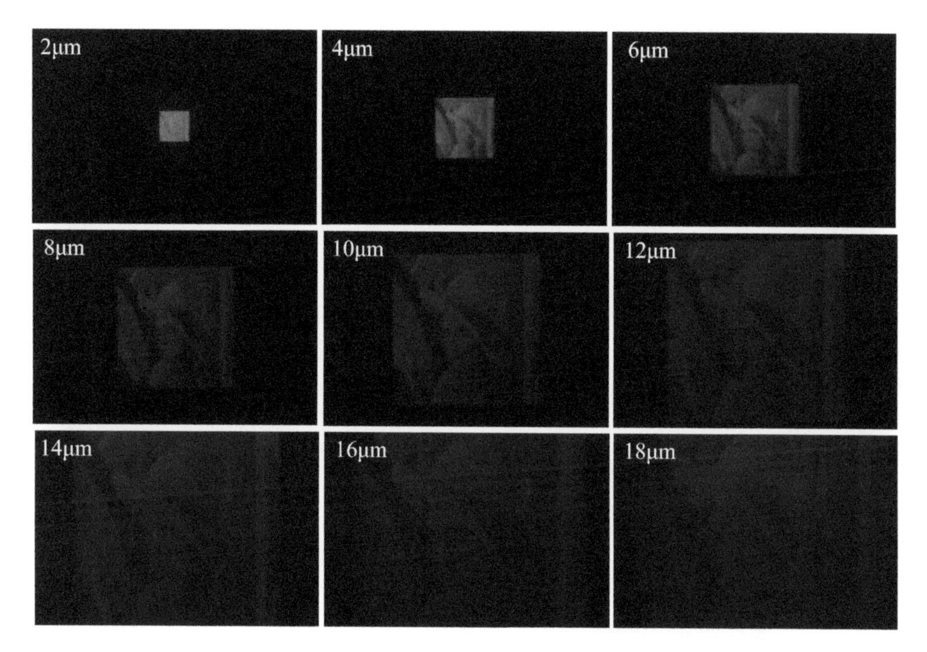

FIGURE 5.20 Zoomable projected images by scale diffraction calculation without using any lens. Reprinted from Ref. [10] with permission, OSA.

information of objects in the measurement field, it is also often used for hologram optimization.

Figure 5.21 shows the GS algorithm for image quality improvement. The GS algorithm optimizes the hologram by imposing constraints on the hologram and its numerical reconstructed image and performing the iterative calculation. By repeating calculations, it is possible to gradually obtain a hologram that reproduces the desired projected image.

In Figure 5.21, first, we give a random phase to the original image and obtain complex amplitude on the hologram surface by diffraction calculation (scale diffraction calculation can be used when zooming). We take only the phase (argument) of this complex amplitude to make it a kinoform (constraint condition on the right side of Figure 5.21). Next, a reconstructed image (projected image) is obtained by performing inverse diffraction calculation from this kinoform. The inverse diffraction calculation can be done by inverting the sign of the propagation distance of the diffraction calculation. By replacing the amplitude component of this reconstructed image with the original image (constraint condition on the left side of Figure 5.21), the gradually optimized hologram (kinoform here) can be obtained.

Figure 5.22 shows reconstructed images from holograms optimized by the GS algorithm. The left and center images show the reconstructed images calculated at 5 and 100 iterations, respectively. As the number of iterations

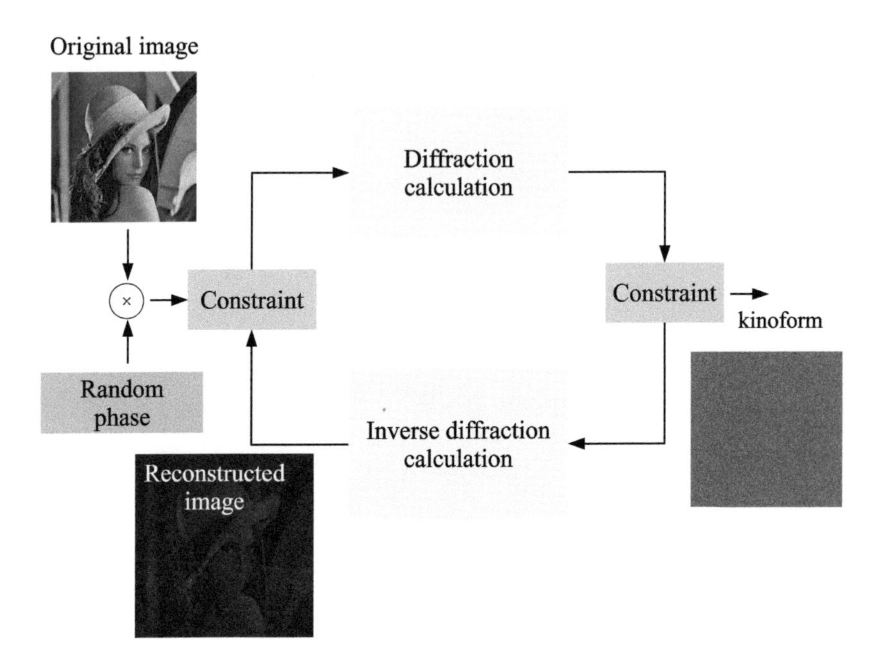

FIGURE 5.21 Hologram optimization using the Gerchberg–Saxton algorithm.

increases, the speckle can be reduced, but it is still noticeable.

As shown on the left of Figure 5.23, a method of calculating a number of holograms having different random phases and averaging the speckle noise by time-sequentially switching these holograms at high speed has been proposed [121]. This method is called the **multi-random phase method**.

In this method, projected images from each hologram have different speckle noise. Therefore, if we use a display device capable of high-speed switching the speckle noise is averaged due to the afterimage effect of the eye and becomes inconspicuous.

Furthermore, speckle noise can be suppressed more effectively by optimizing each hologram with the GS algorithm as shown on the right of Figure 5.23. The reconstructed image on the right side of Figure 5.22 is a projected image when the GS algorithm (5 iterations) and multi-random phase (20 images) are used together. The speckle noise is almost inconspicuous. The image quality of this projected image can be improved, compared with the case of using 100 holograms as a single multi-random phase method.

5.3.4 RANDOM PHASE-FREE METHOD

The **random phase-free method** provides a simple and computationally inexpensive method for enhancing the spatial resolution and reducing the speckle noise of reconstructed images. Calculation of holograms from two-

FIGURE 5.22 Reconstructed images from holograms optimized by the GS algorithm. In the left and the center, the numbers of iterations are 5 and 100. The right figure shows the reconstructed image when combining the GS algorithm (5 iterations) and the multi random phase (20 holograms). Reprinted from Ref. [10] with permission, OSA.

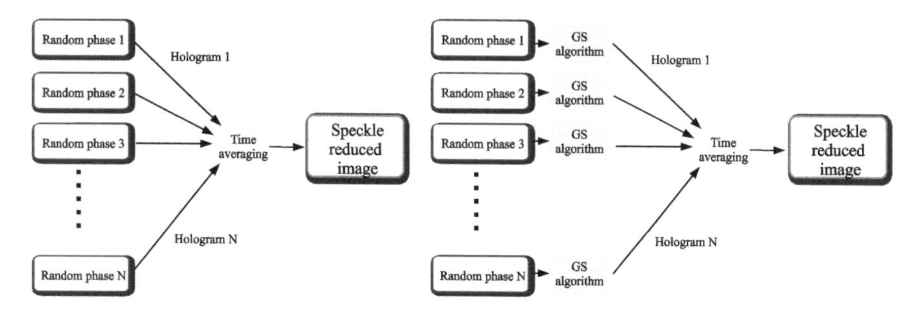

FIGURE 5.23 Combination method of multi-random phase method and the GS algorithm.

and three-dimensional objects generally requires the use of a random phase for widely spreading object light that mainly comprises low-frequency components corresponding to the narrow-angle object light.

Figure 5.24 shows an original image (Figure 5.24(a)) and the image reconstructed from the holograms generated from the original image without (Figure 5.24(b)) and with (Figure 5.24(c)) the random phase. The inspection of Figure 5.24(b) shows that it is nearly impossible for the reconstructed image to retrieve the original image because the object light cannot be recorded on the hologram due to the narrow spreading object light. In contrast, as shown in Figure 5.24(c), the image reconstructed from the hologram generated from the object with the application of the random phase retrieves the original image well. Therefore, the random phase technique has been used in computer holography since the 1960s [122]; unfortunately, the use of the random phase leads to a considerable amount of speckle noise in the reconstructed image.

While this problem can be remedied using iterative optimization methods such as the Gerchberg–Saxton (GS) algorithm to reduce the speckle noise and improve the spatial resolution, these methods are time-consuming. Therefore, for the current computationally efficient methods, the object light processed by the random phase method is uncontrollable, leading to speckle noise and degradation of the spatial resolution of the reconstructed image and motivating the development of new methods for obtaining high-quality images via computer holography.

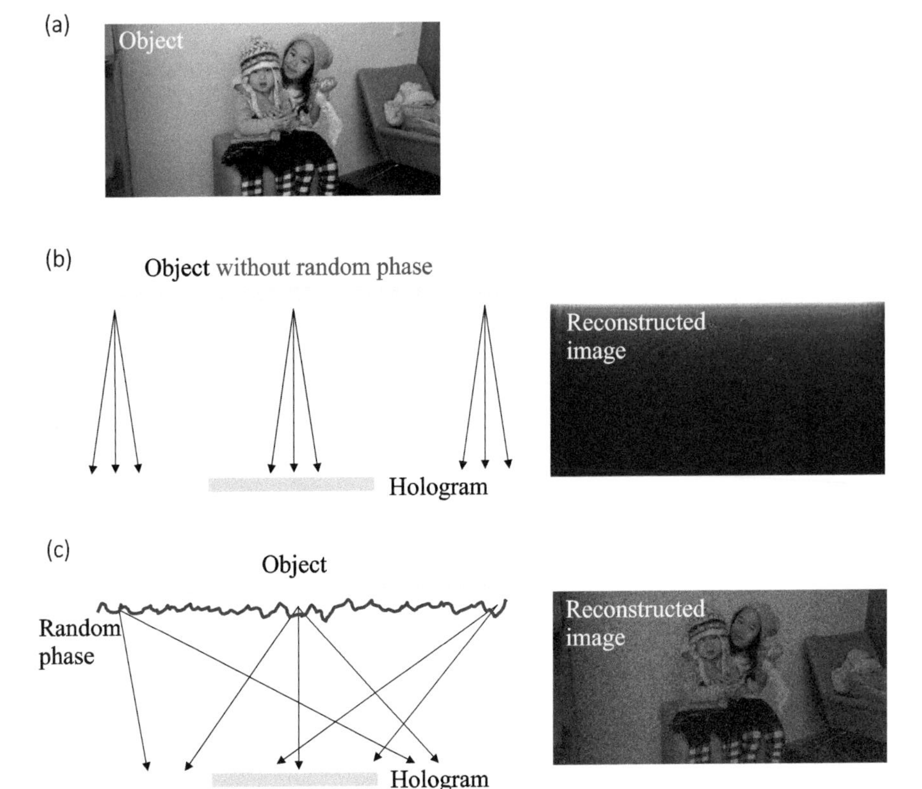

FIGURE 5.24 Reconstructed images from holograms generated from the original image with and without the random phase: (a) original image (b) reconstruction without random phase and (c) reconstruction with random phase. Reprinted from SPIE Newsroom (17 August 2016, DOI:10.1117/2.1201608.006621) with permission, SPIE.

Recently, a simple and computationally inexpensive method called the **random phase-free method** was proposed [87, 123–125]. It drastically reduces speckle noise and enhances the image quality by multiplying the object light by the virtual convergence light. The effectiveness of this method is illustrated in Figure 5.25(a). As can been seen in Figure 5.25(a), the speckle noise of the reconstructed image is lower than those of the images in Figures 5.24(b) and (c). Additionally, as shown in Figure 5.26, we use the USAF1951 test target as the object for examining the spatial resolution. The spatial resolution of the proposed method (Figure 5.26(b)) is clearly better than that of the random phase method (Figure 5.26(a)).

(a)

(b)

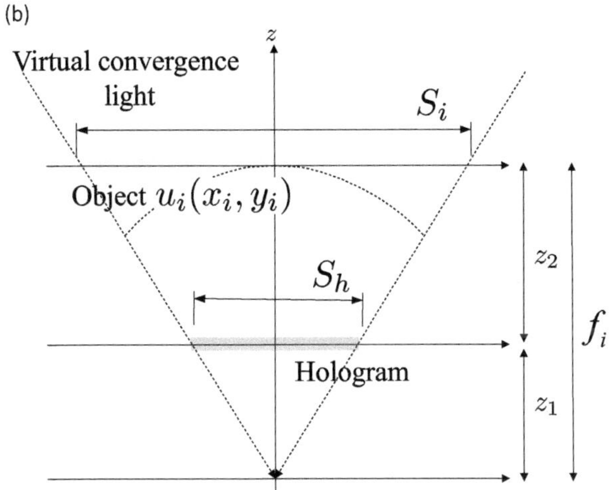

FIGURE 5.25 Reconstructed image from a hologram generated using the proposed method: (a) reconstructed image and (b) calculation setup. Reprinted from SPIE Newsroom (17 August 2016, DOI:10.1117/2.1201608.006621) with permission, SPIE.

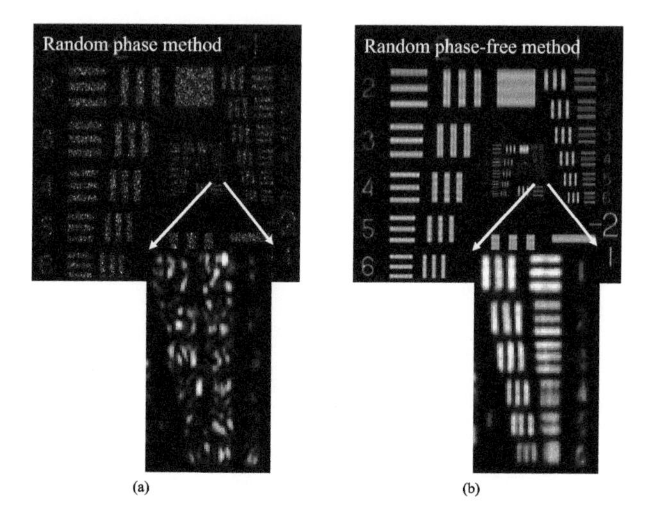

FIGURE 5.26 Comparison of spatial resolution for images obtained by (a) the random phase method and (b) the proposed method. Reprinted from SPIE Newsroom (17 August 2016, DOI:10.1117/2.1201608.006621) with permission, SPIE.

The random phase-free method comprises the multiplication of the object by the virtual convergence light, followed by the calculation of numerical diffraction from the object. Figure 5.25(b) shows the calculation setup of the proposed method. The complex amplitude on the image plane $u_i(x_i, y_i)$ is multiplied using the convergence light given by the following equation:

$$w(x_i, y_i) = \exp\left(-i\frac{\pi(x_i^2 + y_i^2)}{\lambda f_i}\right), \tag{5.38}$$

where λ is the wavelength, $f_i = z_1 + z_2$ is the focal length, z_1 is the distance between the focus point of the convergence light and the hologram that is set to the distance at which the hologram just fits in the cone of the convergence light, and z_2 is the distance between the object and the hologram. The value of f_i is derived from a simple geometric relation, $S_h/2 : S_i/2 = z_1 : f_i$, where the areas of the image and the hologram are given by $S_i \times S_i$ and $S_h \times S_h$, respectively.

The intuitive reason [126] for the effectiveness of the convergence light is outlined in Figure 5.27. As shown in Figure 5.27(a), the convergence light can change the directions of the 0th–, +1st– and −1st–order wave vectors of an object to the hologram represented by the black solid arrows to the wave vectors of the hologram represented by the red arrows. In other words, the object light is distributed on the whole hologram.

As shown in Figure 5.27(b), whose situation is equivalent to a Fourier hologram, if the focal length of the convergence light is $f_i = z_2$, the 0th-order light of the object is strongly concentrated on a small area of the hologram, as indicated by the blue dashed arrow. This is a well-known problem for Fourier holograms. A high dynamic range of the hologram arises from the concentration. The comparison of the dynamic range for the random phase method, Fourier hologram, and the random phase-free method is shown in Figure 5.28. The effective utilization of the dynamic range is particularly important for the amplitude hologram because the almost amplitude spatial light modulators (SLMs) only have 8-bit dynamic range. On the other hand, in the proposed method, the concentration is dispersed over the entire hologram. For our conditions, the average dynamic range of the proposed method is lower by a factor of 5-10 than the situation illustrated in Figure 5.27(b). This feature is favorable for the low-dynamic-range SLMs.

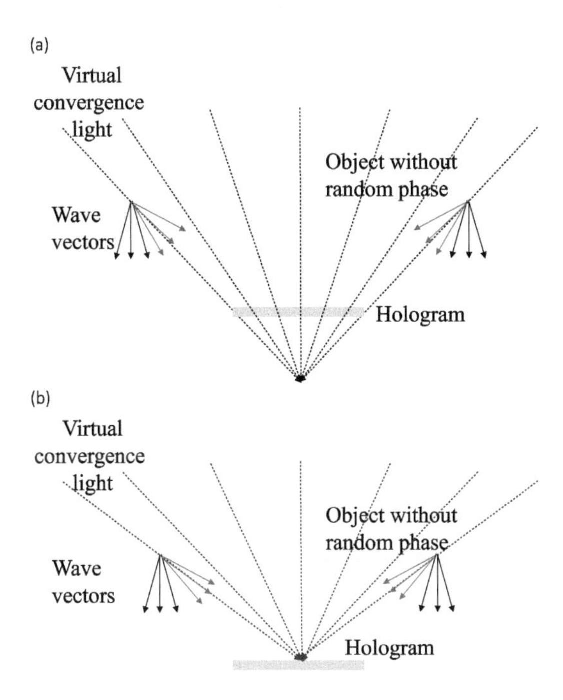

FIGURE 5.27 Intuitive illustration of the effectiveness of convergence light: (a) proposed method and (b) Fourier hologram. Reprinted from SPIE Newsroom (17 August 2016, DOI:10.1117/2.1201608.006621) with permission, SPIE.

The random phase-free method was demonstrated in the application to lensless zoomable holographic projection. Figure 5.29 illustrates the optical reconstructions from the holograms obtained by the random phase and proposed methods. The magnification M values indicated in the figure for the reconstructed images were 1.0, 1.8, and 2.4, respectively. Due to the use of scaled diffraction [29], both reconstructions can be zoomed without using the zoom lens. The optical reconstruction using the random phase-free method is of higher image quality than that obtained using the random phase method

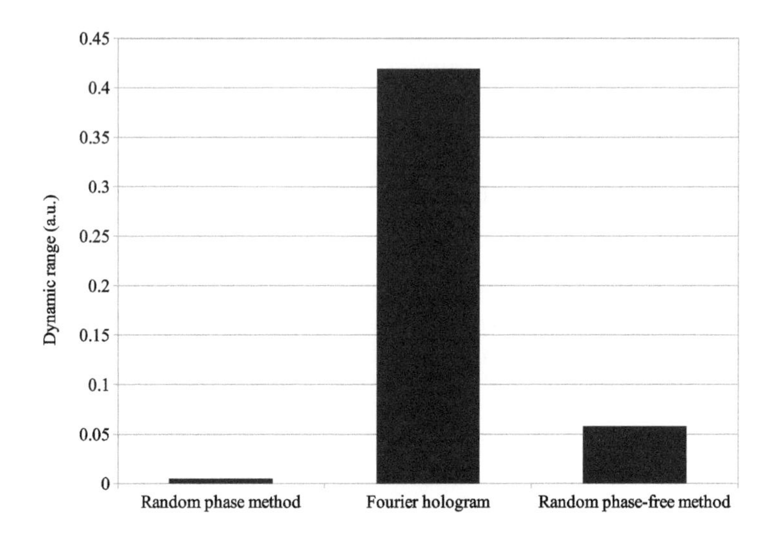

FIGURE 5.28 Comparison of the dynamic range for the random phase method, Fourier hologram, and the random phase-free method.

with lower speckle noise and a sharper image.

In addition, the random phase-free method is effective for color image reconstruction [87]. Figure 5.30 shows the lensless zoomable color reconstructions obtained using the random phase method (left column) and the proposed method (right column). The image obtained by the proposed method is clearly superior, demonstrating the usefulness of the random phase-free method for the reconstruction of color images.

5.4 APPLICATIONS OF DEEP LEARNING TO COMPUTER HOLOGRAPHY

In recent years, much research utilizing **neural networks** (especially **deep learning**) has been conducted in various fields. In the field of image processing, deep learning is applied to image identification and noise elimination, thereby showing performance superior to conventional methods. Also, deep learning takes a lot of time to learn, but once learning the network, there is a feature that can perform fast inference. The application of deep learning is also reported in the field of holography. This chapter introduces application examples of deep learning in holography.

5.4.1 AUTOENCODER-BASED HOLOGRAPHIC IMAGE RESTORATION

As described in Section 5.2, holographic memory [127] is a type of optical memory. In holographic memory, the digital data that is to be recorded

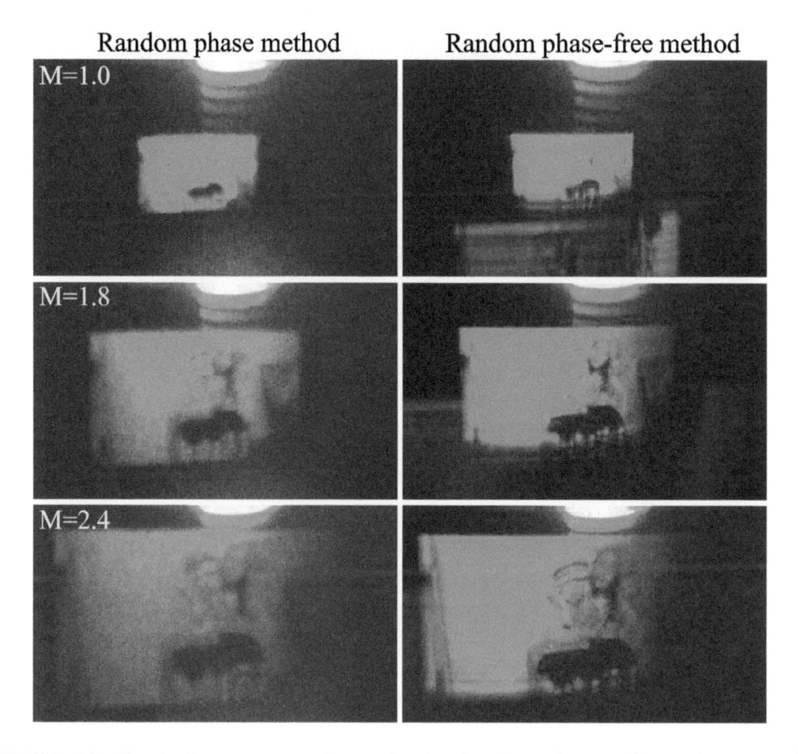

FIGURE 5.29 Optical reconstructions obtained using the random phase method and proposed methods. Reprinted from SPIE Newsroom (17 August 2016, DOI:10.1117/2.1201608.006621) with permission, SPIE.

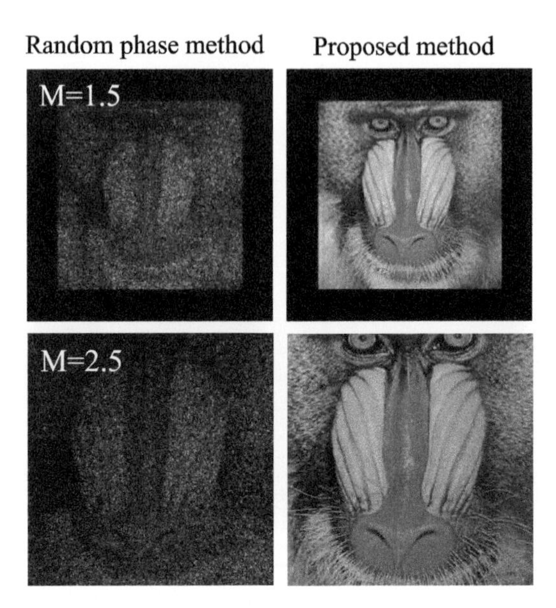

FIGURE 5.30 Lensless zoomable color reconstructions obtained using random phase methods (left column) and proposed method (right column). Reprinted from SPIE Newsroom (17 August 2016, DOI:10.1117/2.1201608.006621) with permission, SPIE.

is converted into a two-dimensional pattern (data page). This data page is recorded onto a recording medium, such as a volume hologram. Holographic memory has a fast access speed because data can be read and written as a two-dimensional data page. Moreover, multiple data pages can be recorded in the same recording area because of the multiplex recording characteristics of holograms. Furthermore, because holograms can be redundantly recorded, even if the recording medium is slightly missing, the recorded data should be able to restore. However, data pages reconstructed from a holographic memory are often degraded by noise. This degradation may make it difficult to recover the recorded digital data.

Recently, using **optical encryption** and optical information hiding has attracted attention because information security has become increasingly important. For example, as shown in Figure 5.31, double random-phase encryption [128, 129] encrypts an image by using two random-phase plates (i.e., the encryption key), and this image can be decrypted using the same key. Figure 5.31(a) shows the optical encryption using two Fourier lenses. The encryption (and decryption) key is the random phase 2. Figure 5.31(b) shows the optical encryption without Fourier lenses. The encryption (and decryption) keys are the random phase 2 and the distances of free space propagation, z_1 and z_2.

Although optical encryption methods such as this one provide high-speed processing and a large number of encryption keys that are random-phase

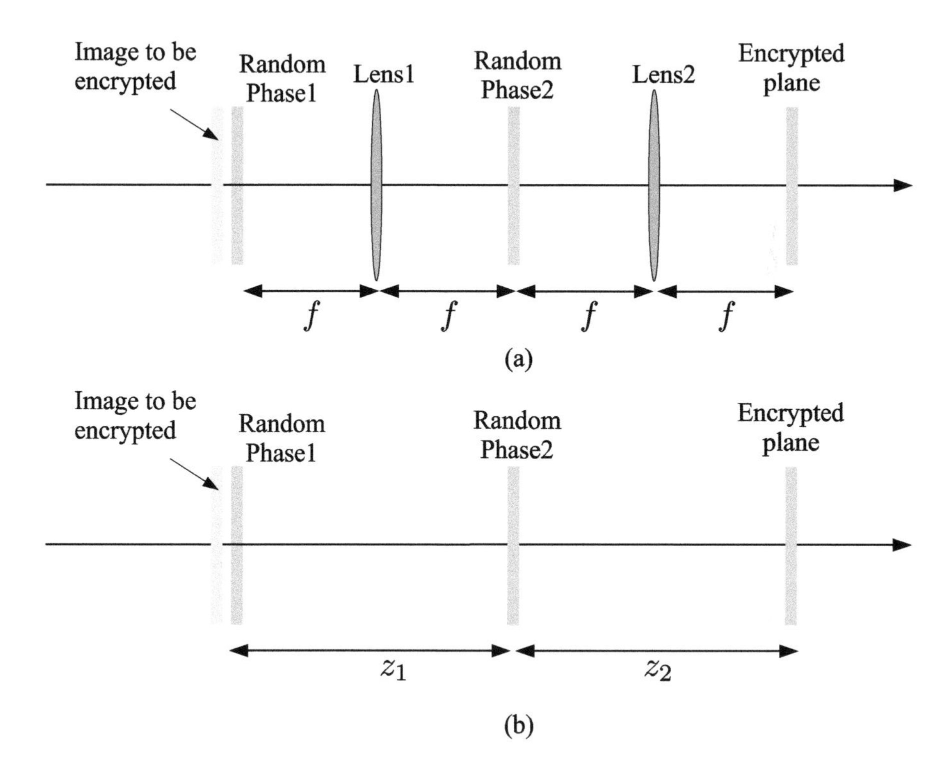

FIGURE 5.31 Double random-phase encryption.

plates, wavelength, and distance, the decrypted images are often contaminated by speckle noise. This means that it may be difficult to discriminate the decrypted images.

To avoid these difficulties, a quick response (QR) code is often used instead of raw images or raw information [130–132]. A QR code is a two-dimensional bar code comprising binary squares. QR codes reconstructed from holograms improve the retrieval of the information. However, due to speckle noise, it is still difficult to recognize reconstructed QR codes in certain recording conditions, which consequently makes retrieving the raw information difficult.

An autoencoder-based holographic image restoration was proposed [11]. An autoencoder [133] is an artificial neural network. The autoencoder has a function that allows it to restore the desired output, even if the input is degraded by noise, which is referred to as a denoising autoencoder [134]. The autoencoder provides a method to restore reconstructed images from holograms that are recorded in a manner similar to the data page used in holographic memory and QR codes.

Figure 5.32 shows the autoencoder-based holographic image restoration method. The reconstructed images are obtained by a numerical diffraction calculation or by optical reconstruction from the holograms. Although the reconstructed images are contaminated by speckle noise, the autoencoder can restore clearer reconstructed images.

In order to train an autoencoder, we need to prepare a large number of training datasets with both original images and the reconstructed images, which are degraded by noise, from the holograms that recorded the original images.

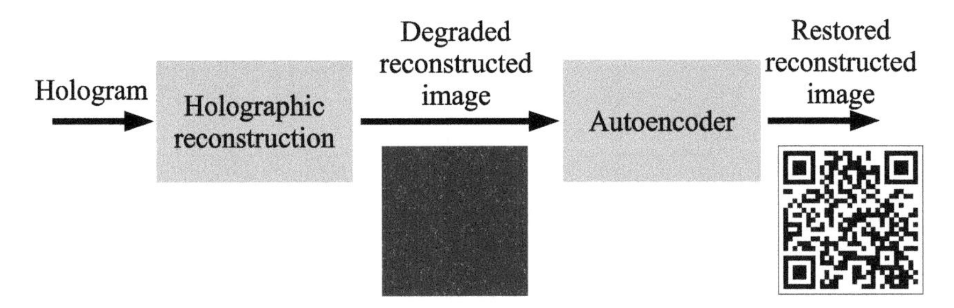

FIGURE 5.32 Autoencoder-based holographic image restoration. Reprinted from Ref. [11] with permission, OSA.

Figure 5.33 shows the autoencoder; it comprises an input layer, $\boldsymbol{x}$, a hidden layer, $\boldsymbol{h}$ and an output layer, $\boldsymbol{o}$, which is used to restore the holographic reconstruction.

The original images and reconstructed images are divided into sub-patterns with $N \times N$ pixels, as shown in Figure 5.33(a). The autoencoder restores each sub-pattern. In order to restore the whole degraded image, it is necessary to input all sub-patterns to the autoencoder. Each pixel of the k-th subpattern

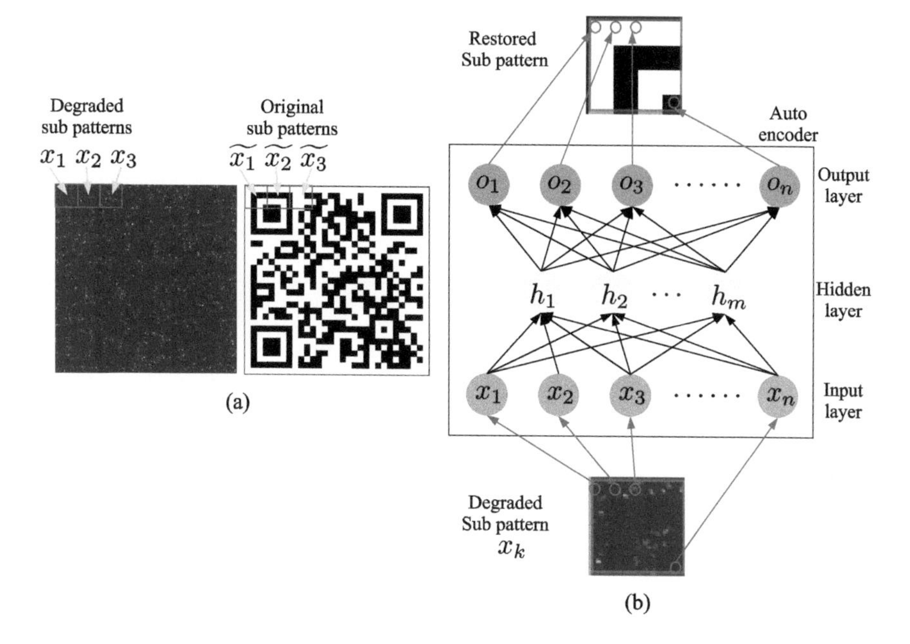

FIGURE 5.33 Autoencoder. Reprinted from Ref. [11] with permission, OSA.

of the degraded and original sub-patterns is vectorized as

$$\boldsymbol{x}_k = [u(\boldsymbol{n}_1), u(\boldsymbol{n}_2), \cdots u(\boldsymbol{n}_{N^2})]^T, \tag{5.39}$$

$$\tilde{\boldsymbol{x}}_k = [\tilde{u}(\boldsymbol{n}_1), \tilde{u}(\boldsymbol{n}_2), \cdots \tilde{u}(\boldsymbol{n}_{N^2})]^T, \tag{5.40}$$

where $\boldsymbol{x}_k \in \mathbb{R}^{N^2 \times 1}$ and $\tilde{\boldsymbol{x}}_k \in \mathbb{R}^{N^2 \times 1}$. u and $\tilde{u}$ are pixels included in the sub-patterns.

As shown in Figure 5.33(b), $\boldsymbol{x}_k$ is the input to the input layer, $\boldsymbol{x}$. The hidden layer, $\boldsymbol{h} \in \mathbb{R}^{M \times 1}$, is calculated using

$$\boldsymbol{h} = f(\boldsymbol{W}\boldsymbol{x} + \boldsymbol{b}), \tag{5.41}$$

where $f(\cdot)$ is an element-wise activation function, $\boldsymbol{W} \in \mathbb{R}^{M \times N^2}$ is the parameter matrix between the input and hidden layers, and $\boldsymbol{b} \in \mathbb{R}^{M \times 1}$ is the bias vector. The rectified linear unit (ReLU) function is used as the activation function.

The output layer, $\boldsymbol{o} \in \mathbb{R}^{N^2 \times 1}$, is calculated by

$$\boldsymbol{o} = f(\widetilde{\boldsymbol{W}}\boldsymbol{h} + \tilde{\boldsymbol{b}}), \tag{5.42}$$

where $\widetilde{\boldsymbol{W}} \in \mathbb{R}^{N^2 \times M}$ is the parameter matrix between the hidden and output layers, and $\tilde{\boldsymbol{b}} \in \mathbb{R}^{N^2 \times 1}$ is the bias vector. Substituting Eq.(5.41) into Eq.(5.42), we can obtain

$$\boldsymbol{o} = f(\widetilde{\boldsymbol{W}} f(\boldsymbol{W}\boldsymbol{x} + \boldsymbol{b}) + \tilde{\boldsymbol{b}}). \tag{5.43}$$

TABLE 5.1
Calculation conditions.

Number of pixels	1,000 × 1,000
Propagation distance	0.05 m
Sampling pitch	4 μm
Wavelength	633 nm

In the autoencoder's training process, a large number of datasets with the original and reconstructed sub-patterns is used to find the parameters ($\boldsymbol{W}$ and $\widetilde{\boldsymbol{W}}$) and biases ($\boldsymbol{b}$ and $\tilde{\boldsymbol{b}}$) that can minimize the loss function, e (least-squares error), for all of $\tilde{\boldsymbol{x}}_k$ and $\boldsymbol{o}$. The loss function is defined as

$$e = |\tilde{\boldsymbol{x}} - \boldsymbol{o}|^2. \tag{5.44}$$

Therefore, we optimize the following equation

$$\underset{\boldsymbol{W},\widetilde{\boldsymbol{W}},\boldsymbol{b},\tilde{\boldsymbol{b}}}{\operatorname{argmin}} \sum_k e. \tag{5.45}$$

We used Adam [135], which is a stochastic gradient descent (SGD) method, to minimize Eq.(5.45). This SGD randomly selects B datasets from all of the datasets $\boldsymbol{x}_k$ and $\tilde{\boldsymbol{x}}_k$. B is referred to as the batch size, and we used a batch size of 100. In addition, we used the Dropout method [136] to prevent overfitting in the autoencoder. Dropout randomly disables N_d percent of units during the autoencoder training process; we used $N_d = 0.8\%$.

After the learning process was complete, degraded sub-patterns, $\boldsymbol{x}_k$, were put into the autoencoder; this enabled us to obtain the restored sub-patterns.

RESULT

We will begin this section by showing the restoration of the data page used in holographic memory. Figure 5.34 shows an example of the data page and the reconstructed image obtained from the hologram using the diffraction calculation. The size of the white and black squares in the data page is 10×10 pixels. The simulation conditions are shown in Table 5.1. We used Chainer [137] as a deep learning framework. The reconstructed image in Figure 5.34 is contrast-enhanced because the contrast was dark. This darkness was due to a high amount of speckle noise contamination. Raw reconstructed images were used in the training process of the autoencoder.

A restored image is obtained from the degraded reconstructed image via the autoencoder. We divide a 1,000 × 1,000-pixel data page into 50 × 50 sub-patterns comprising 20×20 pixels. The on-bit (1) and off-bit (0) are expressed by 10 × 10 pixels, respectively.

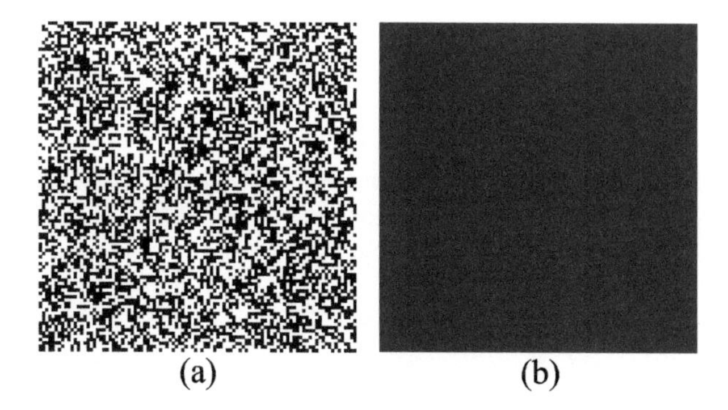

FIGURE 5.34 Examples of data pages: (a) original data page and (b) degraded data page. Reprinted from Ref. [11] with permission, OSA.

The sub-patterns of the data page and degraded sub-pattern are vectorized to $\tilde{\boldsymbol{x}}$ and $\boldsymbol{x}$ with 400 elements, respectively. Therefore, the number of units in the input and output layers is 400. The number of units in the hidden layer is set to 50, which was empirically decided upon.

Figure 5.35 shows the restored image using an autoencoder that had been trained by a dataset comprising 247,500 subpatterns of the data pages and reconstructed images. Figure 5.35(a) is the original data page, while Figure 5.35(b) is the data page restored by the autoencoder after it was trained by 40 iterations of the SGD during the training process. Figure 5.35(c) is the difference between the images in Figures 5.35(a) and 5.35(b). In Figure 5.35(a), the number of original square dots is 10,000. The number of wrong restored binary squares in Figure 5.35(c) is 3. That is, the error rate is 0.03%.

Figure 5.36 shows the restoration of QR codes. We used the same calculation condition and training dataset that were used to obtain the images in Figure 5.35. Figures 5.36(a), 5.36(b) and 5.36(c) are the original QR codes, the QR codes reconstructed using the diffraction calculation, and the QR codes restored using the autoencoder, respectively. The original QR codes contained four different uniform resource locators (URLs). The reconstructed QR codes in Figure 5.36(b) are contrast enhanced. However, we were unable to recognize all of the reconstructed QR codes using smartphones; this was because of a large amount of speckle noise in the images. Conversely, we were able to readily recognize all of the restored QR codes using our smartphones.

5.4.2 DEEP-LEARNING-BASED CLASSIFICATION FOR HOLOGRAPHIC MEMORY

Although holographic memory has a significant advantage as the next-generation data storage, problems do exist, such as bit errors arising from pixel misalignment and noise. Simple thresholding to a data page detected by an image sensor induces bit errors; therefore, we need sophisticated methods

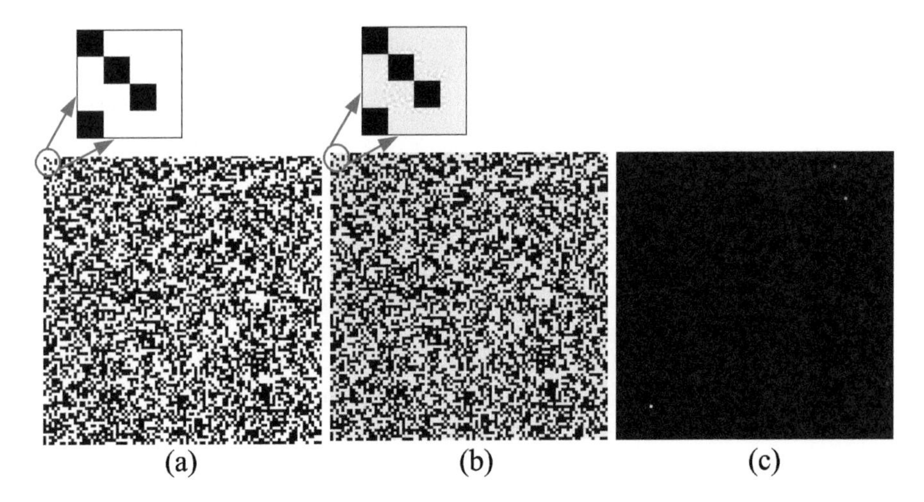

FIGURE 5.35 Restoration of the data page: (a) original data page, (b) data page restored using the autoencoder (40 iterations), and (c) difference between the image of (a) and (b).

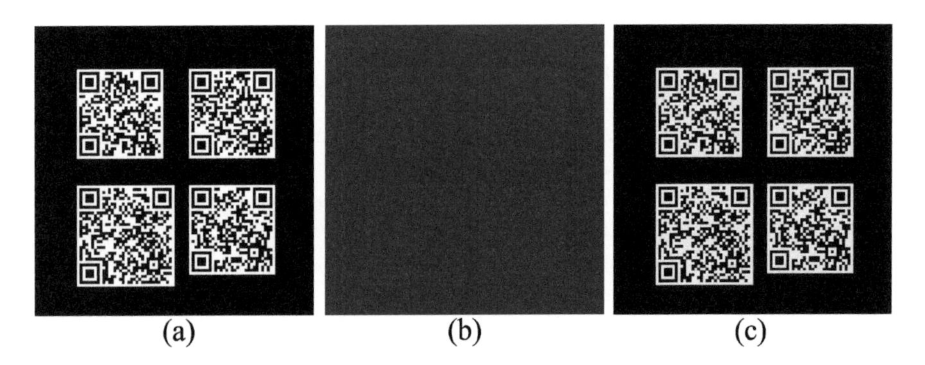

FIGURE 5.36 Restoration of QR codes: (a) original QR codes, (b) reconstructed QR codes, and (c) restored QR codes via the autoencoder. Reprinted from Ref. [11] with permission, OSA.

to correctly detect bits. Precise pixel alignment between reconstructed data pages and an imaging device are required to correctly read data pages. Solutions to this issue have been proposed [138]. Some noises (speckle noise, interpixel interference, interpage interference [139]) contaminate the reconstructed data page. Solutions to the noises have been proposed, such as a Viterbi algorithm [140], a deconvolution method [141], a gradient decent method [142] and an autoencoder [11]. In addition, data pages coded via modulation and error-correction codes help in significantly improving bit errors when compared to raw data pages [143].

Recently, a deep-learning-based classification of data pages was proposed [12]. The deep neural network (**convolutional neural network** [144]) is learned, which is composed of convolutional layers, pooling layer, and fully connected layers, using data pages reconstructed from holograms, and then the learned deep neural network classified data pages. It is significant that the deep neural network can automatically acquire optimum data page classifications from learning datasets without human intervention.

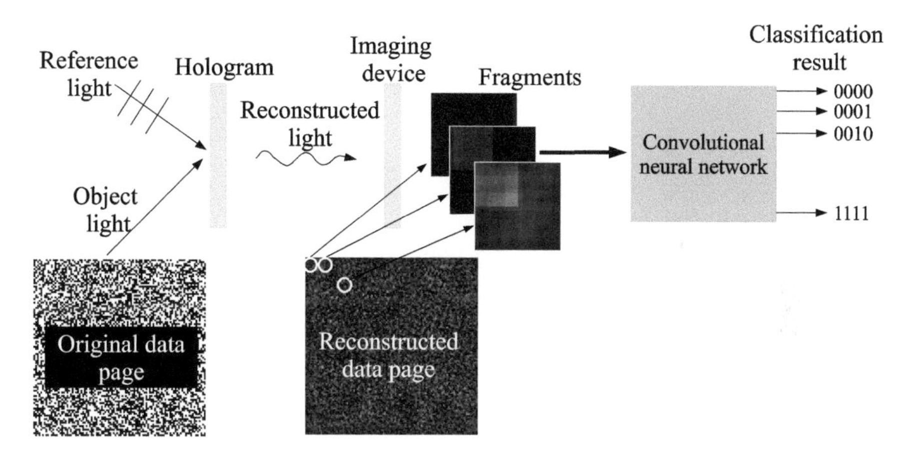

FIGURE 5.37 Deep-learning-based classification for holographic memory. Reprinted from Ref. [12] with permission, OSA.

Figure 5.37 shows the deep-learning-base classification for holographic memory, where data pages are recorded as holograms. These data pages are composed of 4-bit patterns as shown in Figure 5.38[v] Imaging devices such as charge coupled devices (CCDs) captured the reconstructed data pages, which are contaminated by noise, from the holograms. The reconstructed data pages are divided into sub-patterns corresponding to 4-bit original data. The deep neural network in which we used a **convolutional neural network** (**CNN**) classifies sub-patterns into the most similar 4-bit pattern.

[v]We do not use multiple recordings of data pages in the same region of the hologram, or modulation codes such as 6:8 modulation code, or error-correction codes.

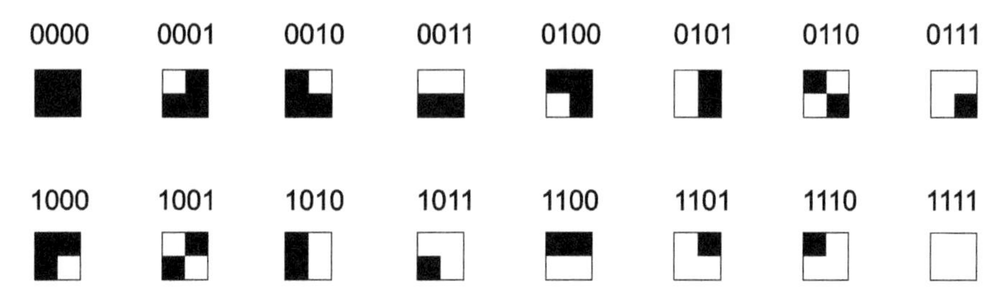

FIGURE 5.38 4-bit representation using 16 sub-patterns. Reprinted from Ref. [12] with permission, OSA.

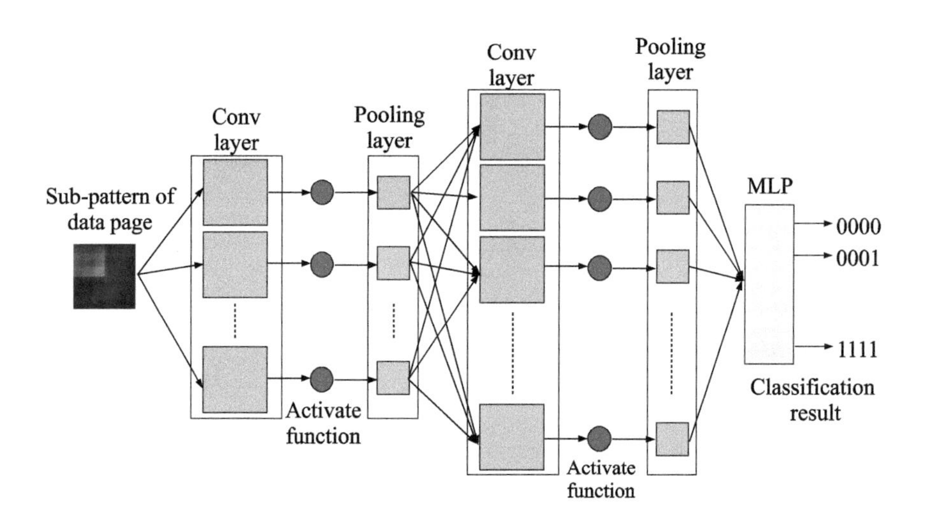

FIGURE 5.39 The structure of the CNN. Reprinted from Ref. [12] with permission, OSA.

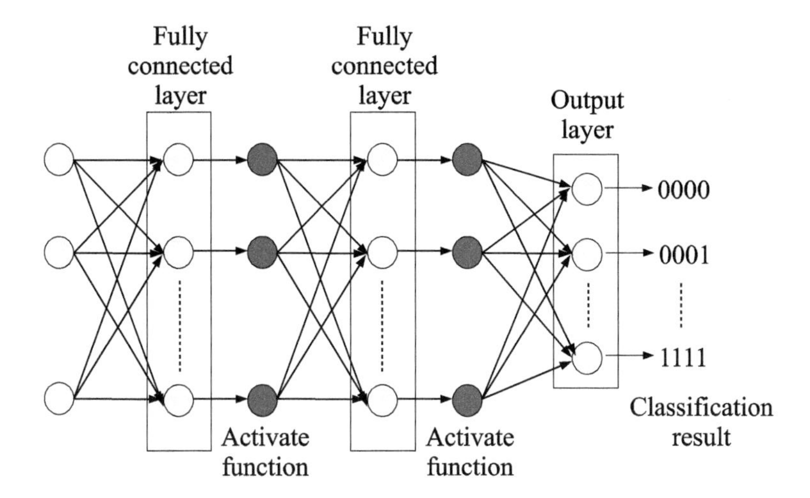

FIGURE 5.40 Multilayer perceptron. Reprinted from Ref. [12] with permission, OSA.

Figure 5.39 shows a CNN for classifying sub-patterns of data pages. The CNN consists of convolution layers, pooling layers, and a multilayer perceptron (MLP) (Figure 5.40) composed of fully connected layers and an output layer.

A convolution layer automatically acquires feature maps of input two-dimensional (2D) data x_{ij}, where the subscript ij denotes the pixel index, using M filters whose filter coefficients denote $h_{pq}^{(m)}$ where $m \in [0, M-1]$. When setting M different filters, the convolution layer acquires M different feature maps. The output of the layer $y_{ij}^{(m)}$ is calculated by

$$y_{ij}^{(m)} = f\left(\sum_{p=0}^{H}\sum_{q=0}^{H} h_{pq}^{(m)} x_{ij} + b_{ij}\right) \tag{5.46}$$

where $f(\cdot)$ is an activate function, H is the filter size and b_{ij} is a bias. The leaky ReLU function ($f(x) = x$ when $x > 0$; otherwise, $f(x) = 0.01x$) is used as the activate function because we confirmed that the classification performance of the activate function was better than that of the ReLU function for our situation.

A pooling layer had the effect of reducing the sensitivity of lateral movement of the input data. In addition, this layer was used for reducing the input data size, resulting in a decrease of the computational complexity. Several pooling layers have been proposed. The max pooling layer that we used was calculated by

$$y_{ij} = \max\{x_{ij}\}. \tag{5.47}$$

Here, this layer divides the input 2D data x_{ij} with $W \times W$ pixels into sub-2D images, and the maximum values in the sub-images are selected and are used to generate the output image y_{ij} with $W/2 \times W/2$ pixels.

The MLP classifies sub-patterns into the most similar 4-bit patterns, as shown in Figure 5.38. The structure of the MLP is shown in Figure 5.40. A fully connected layer in the MLP was calculated by

$$y_j = f\left(\sum_{i=1}^{U} w_{ji}x_i + b_j\right), \tag{5.48}$$

where $f(\cdot)$ is an activate function, U is the number of units, x_i is the input data, w_{ji} is the weight coefficients, y_j is the output data, and b_j is a bias.

In the first and second fully connected layers, we used $U = 128$ and the ReLU function as the activate function. The output layer calculates probabilities of classification using the softmax function, expressed as

$$y_j = \frac{\exp(x_j)}{\sum_{i=1}^{U} \exp(x_i)}, \tag{5.49}$$

where $U = 16$ because we wanted to classify the 16 sub-patterns shown in Figure 5.38.

In the learning process of the CNN, we needed to prepare a large number of datasets composed of reconstructed sub-patterns and corresponding correct answers. Using these datasets, we optimized the parameters (the filter coefficients $h_{pq}^{(m)}$, weight coefficients w_{ji}, and biases) in the CNN. These parameters were optimized by minimizing a cost function. We used the cross-entropy cost function, and we used Adam [135] to minimize the cross-entropy cost function.

RESULTS

We compared the classification performance between the CNN, shown in Figure 5.39, and a conventional MLP, shown in Figure 5.40. The data pages and holographic reconstructions had 1,000 × 1,000 pixels, and the sub-patterns had 20 × 20 pixels. The on-bit (1) and off-bit (0) in a sub-pattern were both expressed by 10 × 10 pixels. Thus, the number of sub-patterns per one data page and holographic reconstruction was 2,500.

We prepared 375,000 fragments (150 data pages × 2,500 sub-patterns) and their holographic reconstructions for the training of the CNN and MLP. For the testing of the CNN and MLP, we used another 125,000 fragments (50 data pages × 2,500 fragments) and their holographic reconstructions. The conditions for the hologram calculation are a wavelength of 633 nm and sampling pitches of the holograms and reconstructions of 4 μm.

We verified the classification performance of the CNN and the conventional MLP when changing the propagation distance z in the hologram generation. Figure 5.41 shows a part of the reconstructed data pages when changing the propagation distance z. We used z=0.05, 0.1 and 0.15 m. As seen, the reconstructed data pages became blurred as the propagation distance increased, resulting in increased difficulty in the classification at the longer distance.

TABLE 5.2
Fragment error rate (FER) when adding the random lateral shift and changing the propagation distance.

	Fragment error rate	
z (m)	MLP	CNN (proposal)
0.05	6.98×10^{-2}	1.44×10^{-4}
0.1	4.23×10^{-2}	4.16×10^{-4}
0.15	2.21×10^{-1}	2.39×10^{-2}

We added a random lateral shift of ± 5 pixels to the reconstructed data pages to verify the robustness against the misalignment between the reconstructed data pages and the image sensor.

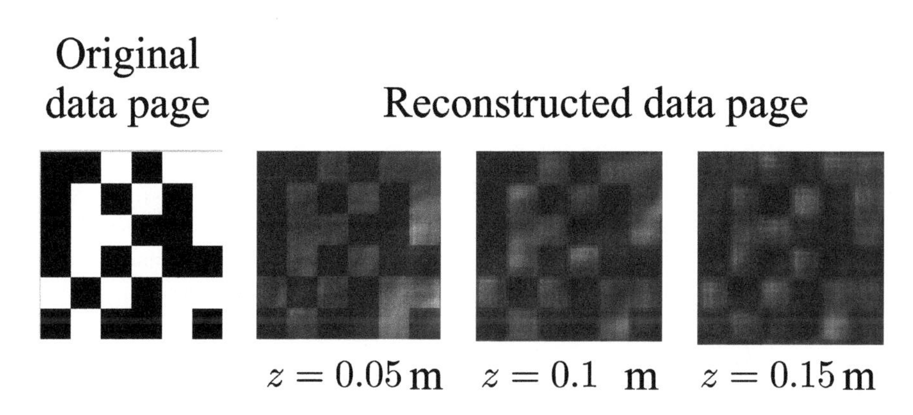

FIGURE 5.41 Examples of reconstructed data pages when changing the propagation distance z. Reprinted from Ref. [12] with permission, OSA.

Table 5.2 shows the accuracy of the classification of the MLP and CNN when adding the random lateral shift and changing the propagation distance. Generally, for the accuracy metrics, in general, a bit error rate (BER) is used in holographic memory; however, we used a fragment error rate (FER) instead of the BER because the MLP and CNN classify the sub-patterns. The FER was calculated by $\text{FER} = N_e/N_t$, where N_e is the number of error sub-patterns and N_t is the number of total sub-patterns in the test process. The FER of the MLP was around 10^{-2}, except at z =0.15 m, whereas the FER of the CNN was around 10^{-4}, except at z =0.15 m, even if the propagation distance was changed. The CNN has an accuracy of two orders of magnitude better than the MLP.

5.4.3 CONVOLUTIONAL NEURAL NETWORK-BASED REGRESSION FOR DEPTH PREDICTION IN DIGITAL HOLOGRAPHY

As described in Chapter 4, digital holography is a promising imaging technique because it enables simultaneously measuring the amplitude and phase of objects, and reconstructing objects in three-dimensional (3D) space from a hologram captured by an imaging sensor [145]. For measuring the objects in 3D space, we need to know the depth position of the recorded object in advance. Subsequently, diffraction is calculated at that position. When using the angular spectrum method [13],[vi] the diffraction calculation is performed by the following equation:

$$u_z(x, y) = \mathcal{F}^{-1}\left[\mathcal{F}\left[u_h(x, y)\right]H(\mu, \nu)\right], \tag{5.50}$$

where $u_h(x, y)$ and $u_z(x, y)$ is a hologram and the complex amplitude in the reconstructed plane at depth z, and the operators $\mathcal{F}[\cdot]$ and $\mathcal{F}^{-1}[\cdot]$ are the Fourier transform and its inverse transform, respectively. $H(\mu, \nu)$ is the transfer function of the angular spectrum method [13].

Without knowing the object's depth position in advance, in previous research, depth search was accomplished by performing multiple diffraction calculations for different depth parameters z, and detecting the most focused depth position by certain focusing metrics.

A number of focusing metrics in digital holography have been proposed; e.g., entropy-based methods [110–112], a Fourier spectrum-based method [113] and a wavelet-based method [114]. Moreover, the Laplacian, variance, gradient, and Tamura coefficients of the intensity of a reconstructed image have been used as focused metrics [8, 113]. Almost all the methods measure the sharpness of the reconstructed intensity images and judge the sharpest image as the focused image. However, such a depth search is time-consuming, because it requires multiple diffraction calculations.

Recently, depth prediction using convolutional neural network-based regression was proposed [146]. This approach can directly predict the depth position with millimeter precision from holograms, without multiple diffraction calculations; that is, this approach estimates the focus position at a high speed. The pioneer work of the depth prediction using deep learning was presented in Refs. [147,148]. The difference between Ref. [146] and Refs. [147,148] is that in Refs. [147, 148], the depth prediction was solved as a classification problem, so the predicted depth becomes a discrete value. On the other hand, since Ref. [146] solves the problem as a multiple regression, the predicted depth becomes a continuous value.

Figure 5.42 shows the network structure of the CNN-based regression. The input layer is for inputting hologram information, where two types of hologram information, the raw interference pattern and the power spectrum of

---

[vi] See Section 2.2.

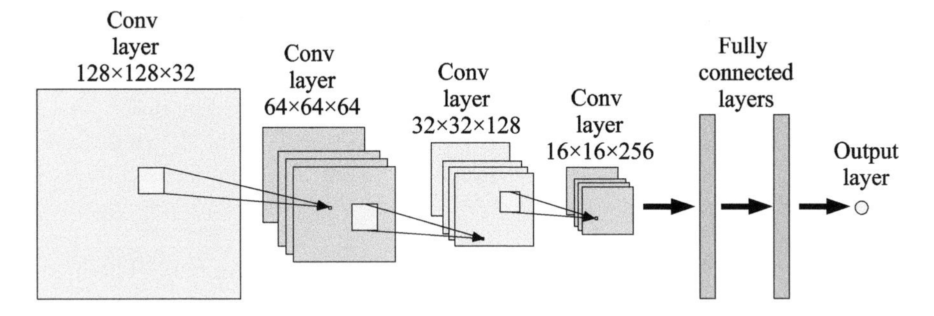

FIGURE 5.42 CNN for predicting focused depth

the hologram, were used. We will compare the differences in the prediction results later.

The convolution layer performs convolution operations with the kernel size of 3×3 pixels to acquire feature maps of the input information. The dimension of the first convolution layer is $128 \times 128 \times 32$ which denotes an input image size of 128×128 pixels and 32 different convolution kernels. All the convolution layers are connected to activation functions (ReLU function) and max-pooling layers. The dimensions of the second, third, and forth convolution layers are $64 \times 64 \times 64$, $32 \times 32 \times 128$ and $16 \times 16 \times 256$. The dimension of each fully connected layer is 2,048. The activation function of the output layer is a linear function (identity function, i.e., $y = x$) because we want to obtain a continuous depth value.

This network is trained by minimizing the loss function, where we use the mean square error (MSE) between the outputs $d_o^{(j)}$ of this network and depth values $d_t^{(j)}$ included in a dataset. The loss function (MSE) is defined as

$$e = \frac{1}{N} \sum_{j=0}^{N-1} |d_o^{(j)} - d_t^{(j)}|^2, \tag{5.51}$$

where the subscript j denotes j-th data in the dataset and N is the size of the dataset. The CNN is trained using Adam [135] with the initial learning rate of 5.0×10^{-4}. The learning rate is automatically decreased when the MSE is stagnated.

RESULTS

The **Tamura metric** is an easier way to find a global solution of depth position compared to the Laplacian, variance, or gradient metrics [8]. The coefficient C_z is calculated as

$$C_z = \frac{\sigma_{I_z}}{\bar{I}_z}, \tag{5.52}$$

where $I_z = |u_z(x, y)|^2$ is the reconstructed intensity, and σ_I and $\bar{I}$ are the standard deviation and average of the intensity, respectively. The Tamura coefficient was used for comparison. The depth search using this metric is performed as follows: we calculate multiple reconstructed intensities I_z using the diffraction calculation equation Eq. (5.50) and varying the depth position z from z_1 to z_2 with a step δz. Subsequently, we calculate the focused metrics C_z for the reconstructed intensities. We can find the focused depth position at the maximum C_z.

We prepare two kinds of datasets as illustrated in Figure 5.43. The holograms and the power spectra of two objects are presented as examples. The first dataset consists of raw images of holograms and their depth values. The hologram size is 1,024 × 1,024 pixels. The reference light with the wavelength of 633 nm is a planar wave. Twenty holograms were captured while moving a same original object along the depth direction ranging from 0.05 m to 0.25 m at random δz intervals of [-5 mm, 5 mm].

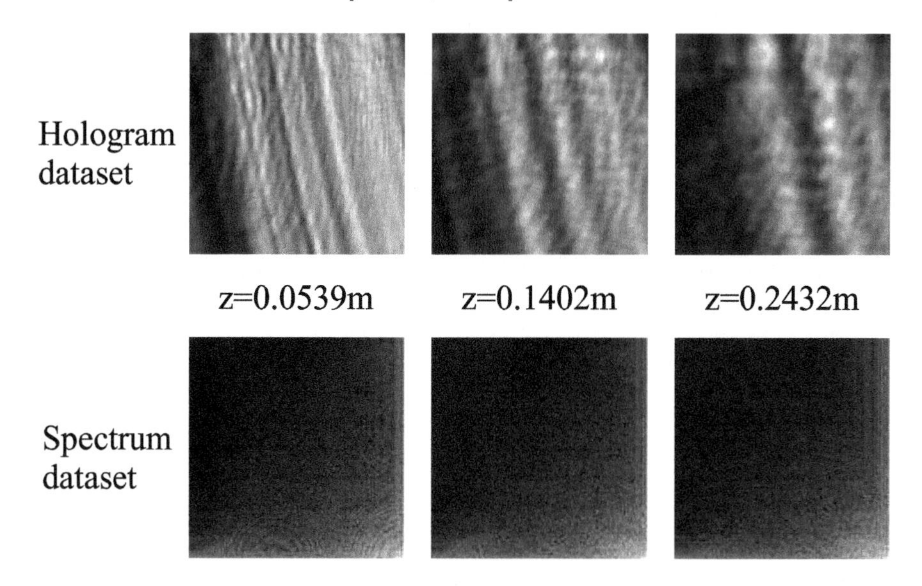

FIGURE 5.43 "Hologram dataset" and "power spectrum dataset." These datasets consist of holograms (or spectra) and the corresponding depth values.

Since the holograms were acquired by an image sensor with 1,024 × 1,024 pixels, we extracted the 128 × 128 pixels in the center of the hologram. The reason for reducing the size of the hologram is to speed up the learning and prediction of the CNN. Accordingly, it helps to simplify the network structure.

The second dataset consists of the power spectra of the holograms and the corresponding depth values, as depicted in Figure 5.43. We call the dataset the "spectrum dataset." The power spectra are calculated from the holograms of 1,024 × 1,024 pixels, and subsequently, the first quadrant of the calculation result is extracted and further reduced to 128 × 128 pixels by linear interpolation. The hologram and spectrum datasets are prepared for training and validation, respectively.

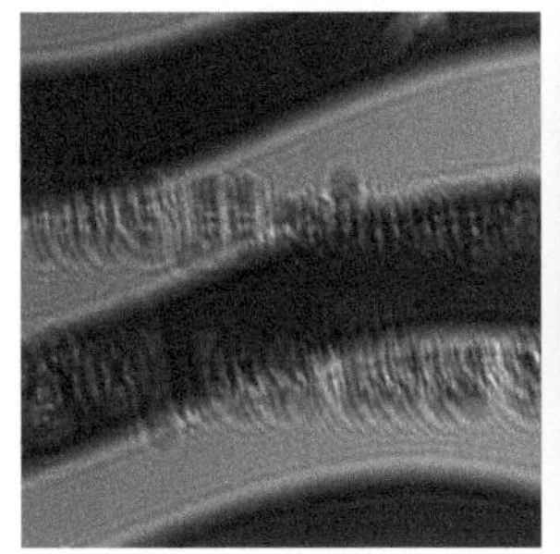
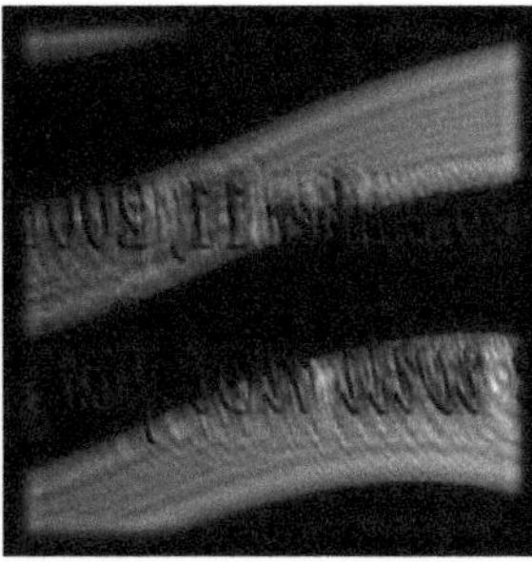
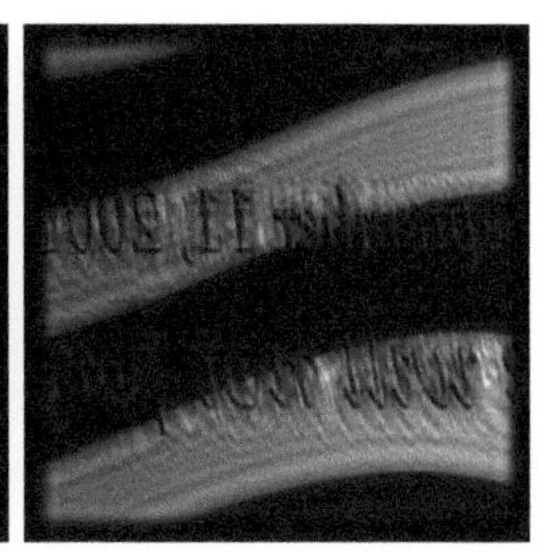

FIGURE 5.44 Hologram and reconstructions: Hologram recorded from an original object at approximately 0.143 m. Reconstructed image of the hologram at $z = 0.147$ m estimated by the maximum value of the Tamura coefficient. The difference between the correct depth and the estimated depth is 4 mm. Reconstructed image of the hologram at $z = 0.138$ m directly predicted by the CNN-based regression. The difference between the correct depth and the predicted depth is 5 mm.

Figure 5.44 (left) depicts a hologram with $1{,}024 \times 1{,}024$ pixels to verify the effectiveness of the CNN. The hologram is recorded from an original object at 0.143 m. We perform the depth search using the Tamura coefficient (Eq. (5.52)) ranging from 0.05 m to 0.25 m with the depth step of 1 mm. The maximum coefficient shows at $z = 0.147$ m.

Subsequently, the CNN can be used to predict the depth value of 0.138 m directly from the power spectrum of the hologram, without the depth search. Figure 5.44 depicts the reconstructed image of the hologram at $z = 0.138$ m. The difference between the actual depth and the predicted depth is 5 mm.

The calculation time of the CNN is 3.5 ms per one power spectrum on an NVIDIA GeForce 970 GTX GPU[vii]; in contrast, the calculation time of the depth search using the Tamura coefficient is 2,892 ms on the same GPU. Compared with the depth search, the CNN can greatly speed up the calculation of depth prediction.

[vii]The detail of the GPU is described in Section 6.2.

6 Hardware implementation

Hardware-based implementations can further boost the computational speed of hologram and diffraction calculations. Typically, **field programmable gate arrays** (**FPGAs**) and **graphics processing units** (**GPUs**) are used for computer holography calculations. This chapter describes how to implement calculations of computer holography in these hardware devices.

6.1 FIELD-PROGRAMMABLE GATE ARRAYS IN COMPUTER HOLOGRAPHY

FPGAs consist of internal memory and many logic blocks in which logic circuits are implemented. A general FPGA structure is shown in Figure 6.1. In the figure, "LB" is a logic block, "SB" is a switch block that can switch the wiring of each block, and "IOB" handles electrical signals outside the FPGA chip.[i]

Logic circuits are designed and written in **hardware description languages** (**HDLs**) such as **VHDL** and **Verilog HDL**. An HDL code is then converted to logic circuit data, which are downloaded to logic blocks and internal memories on an FPGA.

The FPGA design flow is as follows:

1. Designing logic circuits with HDL
2. Logic synthesis
3. Placement and routing
4. Circuit data generation
5. Download circuit data to FPGA

The logic synthesis and place and route correspond to compilation. The logic synthesis replaces an HDL code with a logic circuit. The placement and routing place the gates of the logic circuit on an FPGA. By arranging the placement as close as possible, the transmission distance of the electric signal becomes short, resulting in the circuit operating at high speed.

According to Moore's Law,[ii] the number of logic circuits in FPGAs is increasing year by year. The large-scale digital circuits that had to be implemented in the past with application-specific integrated circuits (ASICs) can also be realized using FPGAs. The advantages of FPGAs are below:

- The circuit can be rewritten any number of times.
- It is suitable for the development of the small lot.
- Circuit operation verification can be done immediately on the spot.

[i] The names LB, IOB, and SB will be changed by FPGA manufacturers.

[ii] The rule of thumb that the number of transistors on an integrated circuit will be doubled every 18 months (= 1.5 years).

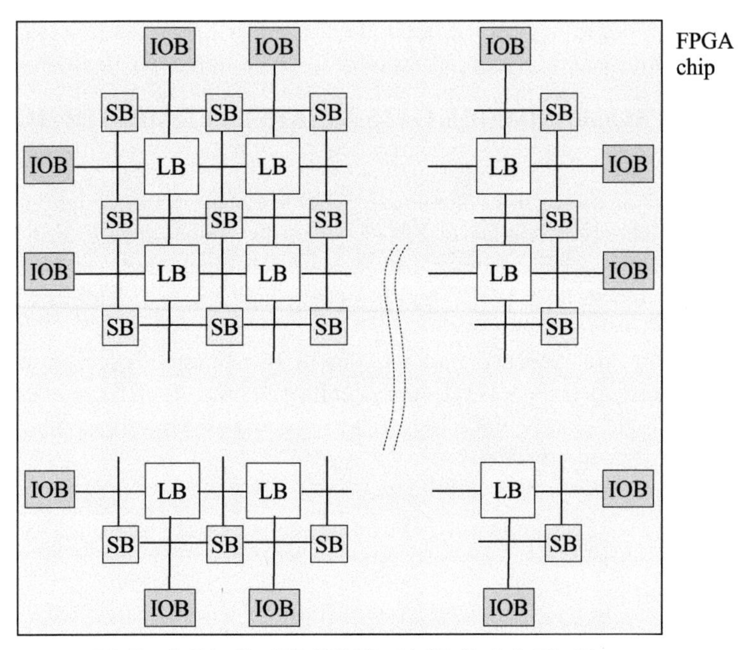

FIGURE 6.1 General FPGA structure: SB (Switch Block), LB (Logic Block) and IOB (I/O Block).

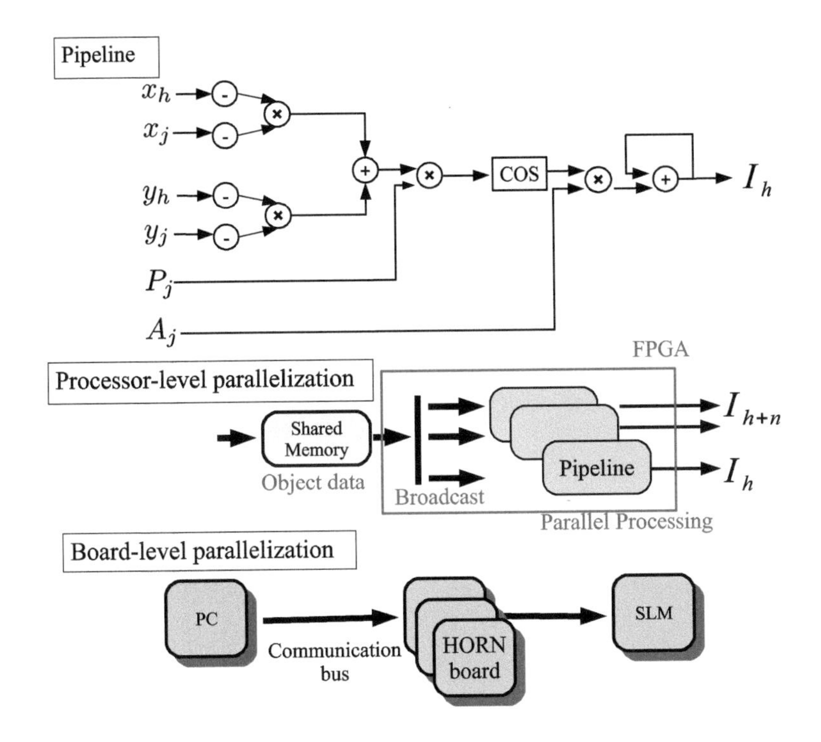

FIGURE 6.2 Block diagram of HORN computers.

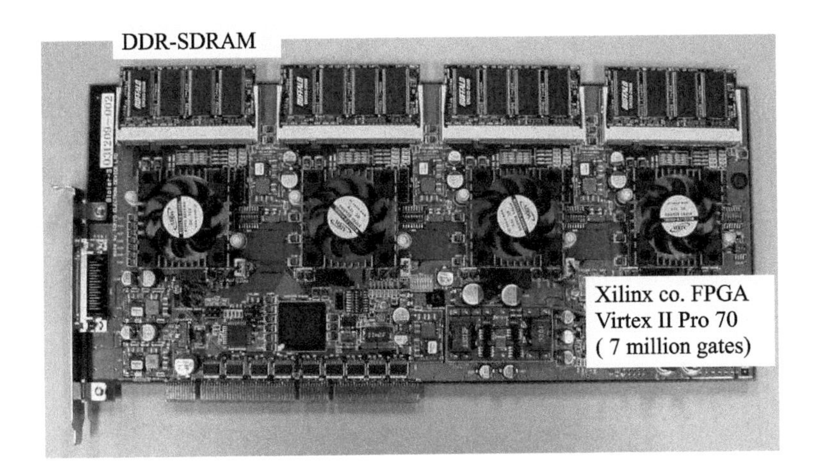

FIGURE 6.3 HORN-5 board.

FPGA-based computers are exceptionally powerful. The problem of the high computational cost of a hologram calculation has been overcome by special-purpose computers for holography named "HOlographic ReconstructioN (**HORN**)," [46, 149–153] that are targeted for use in 3D displays.

In this section, we introduce how to implement a hologram calculation circuit based on HORN-3 [151]. The FPGA computer calculates the following equation.

$$I(x_h, y_h) = \sum_{j=0}^{N-1} \cos\left(\frac{\pi}{\lambda z_j}((x_h - x_j)^2 + (y_h - y_j)^2)\right) \tag{6.1}$$

This calculation is suitable for hardware because it can calculate each pixel of the hologram in parallel.

Figure 6.2 shows a conceptual block diagram of HORN computers, illustrating three levels of HORN parallelization:

- pipeline,
- processor-level parallelization,
- board-level parallelization.

HORN computers designed by pipeline architecture can calculate light intensities on a hologram every clock cycle. The pipeline consists of adders, subtractors, multipliers, LUT for the cosine function and accumulator, and is parallelized in the same FPGA chip. The shared memory that maintains the information (coordinates and light intensities) of 3D object points broadcasts the information to the pipelines enabling the pipelines to calculate light intensities on the hologram simultaneously. Furthermore, FPGA boards on which FPGA chips are mounted, as shown in Figure 6.3, are also parallelized so that many light intensities on the hologram can be calculated at the same time. The calculated light intensities on the hologram are directly outputted to SLMs.

To date, we have constructed six HORN computers, four of which are implemented on FPGA boards except for HORN-1 and HORN-2. HORN computers execute hologram algorithms several thousand times faster than generic computers in those days. Other researches have increased parallelism and improved the memory access of HORN computers in hologram calculations [154, 155].

6.1.1 FIXED-POINT OPERATION

In general, when calculating Eq. (6.1) in a CPU, we use single or double floating-point arithmetic operations. However, when implementing Eq. (6.1) in FPGAs with floating-point arithmetic operations, the circuit size and speed will be large and slow because their circuit structures are complex.

In contrast, the circuit structures of **fixed-point arithmetic** are simpler than those of floating-point arithmetic. A simpler circuit structure reduces

TABLE 6.1
Hologram calculation conditions.

Range of hologram coordinate	$-2048 \leq x_h \leq 2047$
Range of 3D object coordinate	$-2048 < x_j < 2047$, $-2048 < y_j < 2047, z_j \leq 1$ m
Wavelength λ	633 nm
Sampling pitch p	10 μm
Maximum number of object N	512

the circuit size and increases calculation speed (clock cycle). In this section, we describe how to convert the floating-point arithmetic of Eq. (6.1) to fixed-point arithmetic operations.

We assume that the hologram calculation conditions for the hardware implementation are shown in Table 6.1. Under these conditions, let us decide the bit width of the variables in Eq. (6.1). We rewrite Eq. (6.1) to the following equation:

$$I(x_h, y_h) = \sum_{j=0}^{N-1} \cos\left(P_j((x_h - x_j)^2 + (y_h - y_j)^2)\right) \tag{6.2}$$

where P_j is defined as $P_j = \frac{\pi}{\lambda z_j}$.

x_h, y_h, x_j, y_j are integer values ranging from $-2,048$ to $2,047$, so these variables can be expressed by 12-bit fixed-point numbers (i.e., 2^{14}=4,096). Since $P_j > 0$ is always satisfied under the calculation conditions in Table 6.1, we convert it to an unsigned 32-bit fixed point by specifying sf as 1 in the double_to_fixed function as shown in List 6.1.

Listing 6.1 Converting a double precision floating-point number to a fixed-point number.

```
1  long long int double_to_fixed(double data, int sf, int width)
2  {
3    long long int tmp = *(long long int*)(&data);
4    long long int mant = tmp & 0xfffffffffffffLL;
5    unsigned long exp = (tmp >> 52) & 0x7ff;
6    unsigned char sign =(tmp >> 63) & 0x01;
7
8    mant |= 0x10000000000000LL;
9
10   mant >>= 1023 - exp;
11   mant >>= (53 - width - sf);
12   if(sign) mant = mant * -1;
13
14   return mant;
15 }
```

Let us consider converting the cosine function in Eq. (6.2) to a fixed point number (Listing 6.2). Since the amplitude of the cosine takes a range from -1 to $+1$, it is necessary to express positive and negative numbers with a fixed-point number. We use two's complement to express negative numbers. The phase of the cosine function is divided by 256 (i.e., 256 addresses), and the amplitude also divides -1 to $+1$ by 256 (8 bits precision). Since the cosine function is stored in a LUT, the required amount of the memory is 256 addresses $\times$ 8 bits $=$ 256 bytes.

Listing 6.2 Converting the cosine function in Eq. (6.2) to a fixed-point number.

```
1  #define PI 3.14159265358979323846264338327950
2  for(i=0; i<256; i++){
3    double c = cos(2.0 * PI * (double)i / 256.0);
4    unsigned char tmp;
5
6    if(c >= 1.0) tmp = 0x07f;
7    else if(c <= -1.0) tmp = 0x80;
8    else tmp=double_to_fixed(c, 0, 8);
9    cos_tbl[i]=tmp;
10 }
```

An emulator is necessary to check whether the hologram can be calculated with limited precision (fixed-point arithmetic). Generally, as the number of bits of the arithmetic circuit increases, the circuit scale increases, and the operation speed decreases. Therefore, it is important to reduce the number of bits as much as possible by using the emulator.

Listing 6.3 is the source code of the emulator written in C language. This function "pipe" gives the coordinates (x_h, y_h) of the hologram plane, the coordinates and intensity A_j of 3D objects, and the number of the object points N as the arguments. The return value is the light intensity $I(x_h, y_h)$ as a long type.

Listing 6.3 Emulator of hologram calculation in a fixed-point number.

```
1  unsigned short MYABS(short a)
2  {
3    if(a<0) return (~a + 1); //(NOT A) + 1
4    else return a;
5  }
6
7
8  long pipe(short xa, short ya, short *xj, short *yj, unsigned long *pj,
         unsigned char *aj, int N)
9  {
10   long I = 0;
11   int i;
12
13   for(i=0;i<N;i++){
```

```
14      short SUBX_wire = xa−xj[i];
15      short SUBY_wire = ya−yj[i];
16
17      unsigned short ABSX_wire = MYABS(SUBX_wire);
18      unsigned short ABSY_wire = MYABS(SUBY_wire);
19
20      unsigned int SQRX_wire = ABSX_wire * ABSX_wire;
21      unsigned int SQRY_wire = ABSY_wire * ABSY_wire;
22
23      unsigned int ADDXY_wire = SQRX_wire + SQRY_wire;
24
25      long long int MULTPJ_wire=ADDXY_wire * pj[i];
26
27      unsigned char COSIN_wire = (MULTPJ_wire >> 24) & 0xff;
28      char COS_wire = cos_tbl[COSIN_wire];
29
30      short MULTAJ_wire = aj[i] * COS_wire;
31
32      char tmp = (MULTAJ_wire>>8) ;
33      I += (long)tmp;
34    }
35    return I;
36  }
```

The details of each line of the list are described below:

Lines 14 and 15 Under the calculation conditions in Table 6.1, x_a, y_a, x_j and y_j take a range of $-2,048$ to 2,047. To represent this range, we use two's complement number with 12-bit. Unfortunately, there is no 12-bit data type in the C language. Therefore, we use the short type (16-bit length) closest to 12 bits. Subtraction of 12-bit data becomes 13 bits (SUBX_wire and SUBY_wire).

Lines 17 and 18 These lines calculate the absolute values of the subtracted results using MYABS function. The MYABS function converts a negative number to the positive number by using two's complement. The absolute operation reduces SUBX_wire and SUBY with 13 bits to those with 12 bits; therefore, the subsequent square operations become 24 bits. If we do not use the absolute operation, the subsequent square operations become 26 bits. Compared cases with/without using the absolute operation, the bit width can be reduced by 2 bits.

Line 23 The bit width of the added result (ADDXY_wire) is 25 bits.

Line 25 This line multiplies 25-bit ADDXY_wire with 32-bit P_j. The bit width of the result (MULTPJ_wire) is 57 bits. We use the 64-bit integer type (long long int) to store the 57-bit data.

Line 27 This line calculates the address of the COS table. Figure 6.4 shows the decimal points of ADDXY_wire and P_j, and the multiplication result. ADDXY_wire has the decimal point on the rightmost side, and P_j is al-

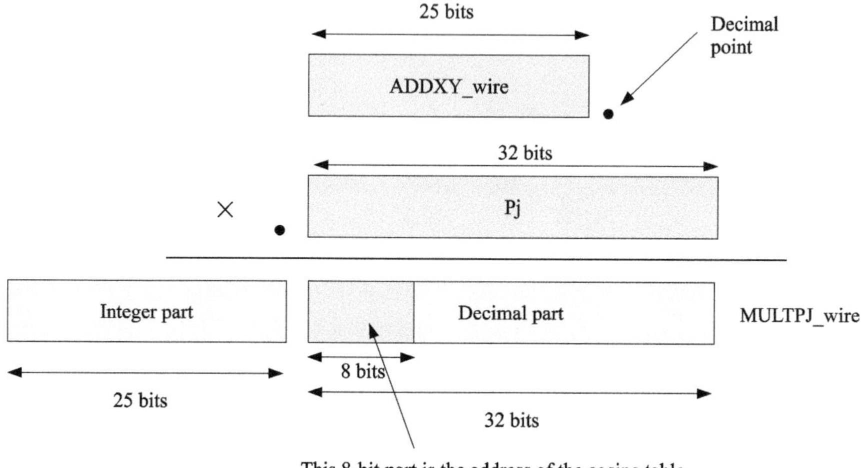

FIGURE 6.4 Decimal points of ADDXY_wire and P_j, and the multiplication result.

ways less than 1, so the decimal point is placed on the leftmost side. The multiplication result, MULTPJ_wire, can be divided into the integer and decimal parts. When inputting the multiplication result in the cosine table, the integer part can be ignored because $\cos(2\pi(MULTPJ\_wire)) = \cos(2\pi(integer + decimal)) = \cos(2\pi \times decimal)$. The cosine table has 256 addresses (= 8 bits), so the upper the 8 bits of the decimal part becomes the address of the cosine table.

Line 28 This line calculates the cosine function using the cosine table. The calculation result (COS_wire) is an 8-bit integer of two's complement representation.

Line 30 This line multiplies the light intensity A_j with COS_wire. Since Aj and COS_wire are 8 bits, the multiplication result (MULTAJ_wire) becomes 16 bits.

Line 32 This line reduces the bit width of MULTAJ_wire by right-shifting MULTAJ_wire to the right to make it 8 bits. Since both the effective digits of A_j with COS_wire are 8 bits, the significant digit of the multiplication result is also 8 bits. Therefore, it takes the upper 8 bits of the multiplication result.

Line 33 This line accumulates tmp. Since we assume the maximum number of object points is 512 (= 9 bits), the bit width of I is enough for 17 bits (8 bits + 9 bits) without causing overflow.

Table 6.2 summarizes the presence or absence of the sign, and the bit width of each operation determined by the emulator. Figure 6.5 shows the calculation results of holograms of two object points using the emulator. This result is almost the same as when using the double precision floating point.

The calculation conditions for this hologram are shown below. The size of the hologram is 1,024 × 768 pixels. The coordinates of the two object points are $(x_j, y_j) = (-256, 0)$ and $(x_j, y_j) = (+256, 0)$. The wavelength is $\lambda = 633$ nm, the sampling pitch is $p = 10$ μm, and the distance between the hologram and the object is 1 m. In this case, P_j is a 32-bit integer and is 00052D36 (in hexadecimal notation).

TABLE 6.2

Presence or absence of sign, and the number of bits required for each variable.

Variable	Sign/Unsign	Bit width
$\mathrm{x_h}$, $\mathrm{y_h}$	Sign	12
$\mathrm{x_j}$, $\mathrm{y_j}$	Sign	12
$\mathrm{P_j}$	Unsign	32
$\mathrm{A_j}$	Unsign	8
SUBX_wire, SUBY_wire	Sign	13
ABSX_wire, ABSY_wire	Unsign	12
SQRX_wire, SQRY_wire	Unsign	24
ADDXY_wire	Unsign	25
ADDPJ_wire	Unsign	57
COSIN_wire	UnSign	8
COS_wire	Sign	8
MULTAJ_wire	Sign	16
tmp	Sign	8
I	Sign	17

6.1.2 HDL IMPLEMENTATION

The structure of the hologram calculation circuit is shown in Figure 6.6. This circuit consists of adders, subtractors, multipliers and a cosine table. The bit width of each arithmetic unit was determined by the emulator (Listing 6.3). 6.2. This circuit performs the calculation of Eq. (6.2) by a pipeline circuit.

Pipeline processing is a method for speeding up a large number of processes that are continuously performed. A typical example of pipeline processing is often found in a factory production line. Figure 6.7 shows an example of calculating $(a + b) \times 5 - 1$ for three different data, $(a_1, b_1), (a_2, b_2), (a_3, b_3)$. The upper part of the figure is non-pipeline processing, and each processing is executed sequentially. On the other hand, the lower part of the figure shows the pipeline processing. This pipeline processing divides $(a + b) \times 5 - 1$ into three stages: the first stage is $(a + b)$, the second stage is $(a + b) \times 5$ and the third stage is $(a + b) \times 5 - 1$. It can perform these stages in parallel by inputting the next data a_n, b_n while processing the previous data, a_{n-1}, b_{n-1}.

The pipeline of Figure 6.6 can be divided into the following seven stages:

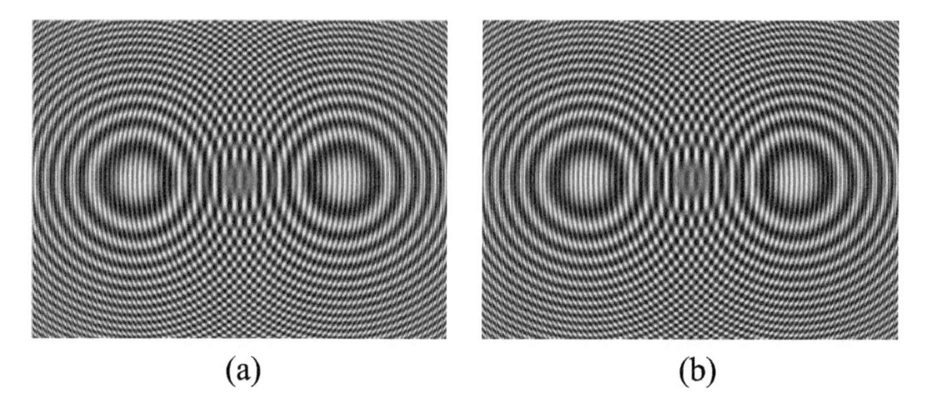

FIGURE 6.5 Holograms of two object points using the emulator.

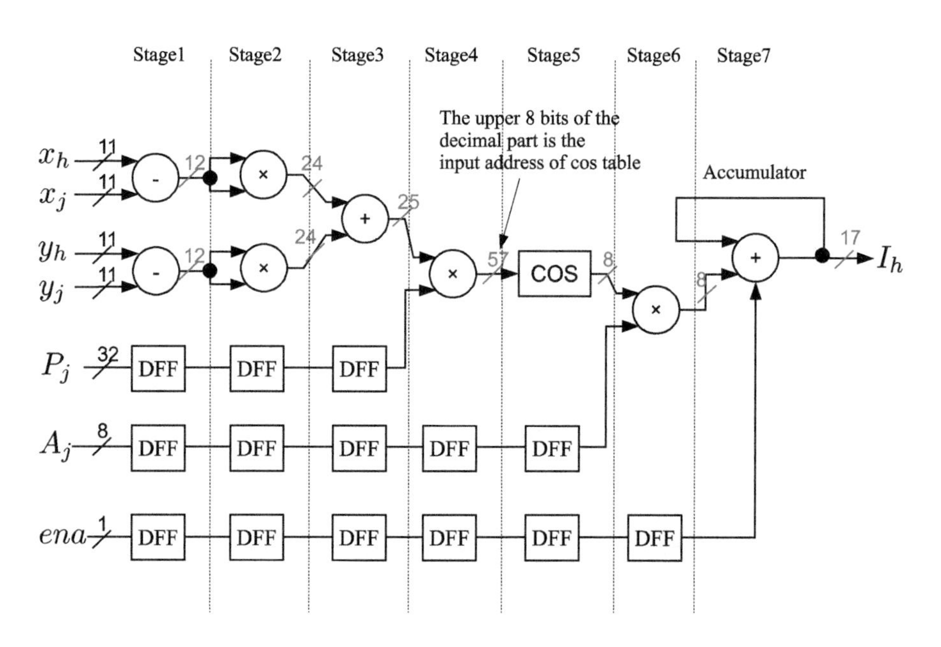

FIGURE 6.6 Pipeline circuit for hologram calculation.

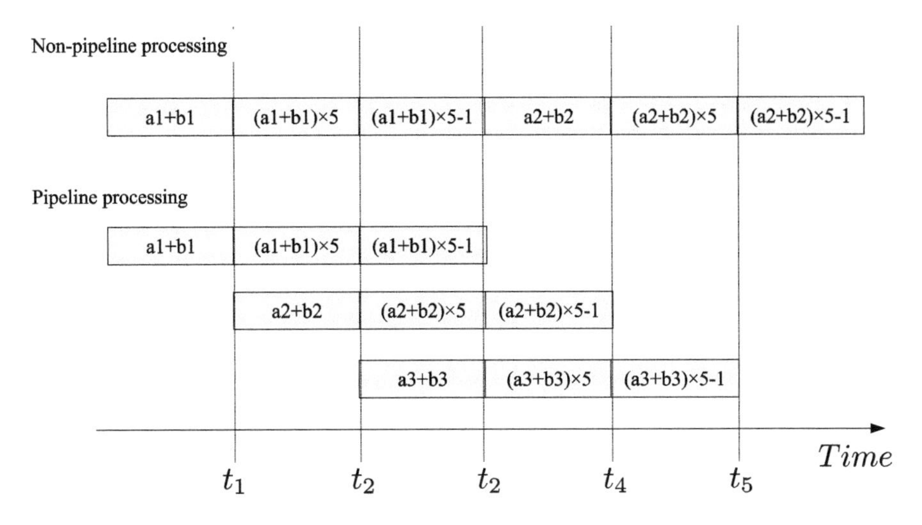

FIGURE 6.7 Example of pipeline processing.

Stage 1: Calculation of $(x_h - x_j)$ and $(y_h - y_j)$ with subtractors.
Stage 2: Calculation of absolute value. Squared calculation of $(x_h - x_j)^2$ and $(y_h - y_j)^2$ by multipliers.
Stage 3: Calculation of $(x_h - x_j)^2 + (y_h - y_j)^2$ with an adder.
Stage 4: Calculate $P_j \times ((x_h - x_j)^2 + (y_h - y_j)^2)$ with a multiplier.
Stage 5: Calculate $\cos(P_j \times (x_h - x_j)^2 + (y_h - y_j)^2)$ in the cosine table.
Stage 6: Calculate $A_j \times \cos(P_j \times (x_h - x_j)^2 + (y_h - y_j)^2)$ with a multiplier.
Stage 7 Calculate $\sum A_j \times \cos(P_j((x_h - x_j)^2 + (y_h - y_j)^2))$ with an accumulator.

In Figure 6.6, "ena" is the control signal.[iii] When "ena" is high level, the accumulator starts the operation. "DFF" represents a D flip-flop. It constructs the pipeline register[iv] by connecting the D-FFs in cascade, and supplies data to the arithmetic units when A_j, P_j is required.

The VHDL source code for the pipeline of Figure 6.6 is shown in List 6.4.

Listing 6.4 VHDL source code for the pipeline of Figure 6.6.

```
1 library ieee;
2 use ieee.std_logic_1164.all;
3 use ieee.std_logic_unsigned.all;
4
5 entity pipeline is
6 end pipeline;
```

[iii] Enable signal.

[iv] Shift register.

```
 7 architecture rtl of pipeline is
 8
 9 signal CLK : std_logic;
10 signal R : std_logic;
11
12 signal xj : std_logic_vector(11 downto 0);
13 signal xh : std_logic_vector(11 downto 0);
14 signal yj : std_logic_vector(11 downto 0);
15 signal yh : std_logic_vector(11 downto 0);
16
17 signal SUBX_wire : std_logic_vector(12 downto 0);
18 signal SUBY_wire : std_logic_vector(12 downto 0);
19
20 signal SQRX_wire : std_logic_vector(23 downto 0);
21 signal SQRY_wire : std_logic_vector(23 downto 0);
22
23 signal ADDXY_wire : std_logic_vector(24 downto 0);
24 signal pj : std_logic_vector(31 downto 0);
25 signal MULTPJ_wire : std_logic_vector(56 downto 0);
26 signal COS_wire : std_logic_vector(7 downto 0);
27 signal aj : std_logic_vector(7 downto 0);
28 signal MULTAJ_wire : std_logic_vector(15 downto 0);
29 signal I : std_logic_vector(16 downto 0);
30 signal ena : std_logic;
31
32 subtype WAVE is std_logic_vector (7 downto 0);
33 type ROM is array (0 to 255) of WAVE;
34 constant cos_tbl : ROM := (
35   X"7F",X"7F",X"7F",X"7F",X"7F",X"7F",X"7E",X"7E",
36   X"7D",X"7C",X"7C",X"7B",X"7A",X"79",X"78",X"77",
37   X"76",X"75",X"73",X"72",X"70",X"6F",X"6D",X"6C",
38   X"6A",X"68",X"66",X"64",X"62",X"60",X"5E",X"5C",
39   X"5A",X"58",X"55",X"53",X"51",X"4E",X"4C",X"49",
40   X"47",X"44",X"41",X"3F",X"3C",X"39",X"36",X"33",
41   X"30",X"2E",X"2B",X"28",X"25",X"22",X"1F",X"1C",
42   X"18",X"15",X"12",X"0F",X"0C",X"09",X"06",X"03",
43   X"00",X"FD",X"FA",X"F7",X"F4",X"F1",X"EE",X"EB",
44   X"E8",X"E4",X"E1",X"DE",X"DB",X"D8",X"D5",X"D2",
45   X"D0",X"CD",X"CA",X"C7",X"C4",X"C1",X"BF",X"BC",
46   X"B9",X"B7",X"B4",X"B2",X"AF",X"AD",X"AB",X"A8",
47   X"A6",X"A4",X"A2",X"A0",X"9E",X"9C",X"9A",X"98",
48   X"96",X"94",X"93",X"91",X"90",X"8E",X"8D",X"8B",
49   X"8A",X"89",X"88",X"87",X"86",X"85",X"84",X"84",
50   X"83",X"82",X"82",X"81",X"81",X"81",X"81",X"81",
51   X"80",X"81",X"81",X"81",X"81",X"81",X"82",X"82",
52   X"83",X"84",X"84",X"85",X"86",X"87",X"88",X"89",
53   X"8A",X"8B",X"8D",X"8E",X"90",X"91",X"93",X"94",
54   X"96",X"98",X"9A",X"9C",X"9E",X"A0",X"A2",X"A4",
```

```
55  X"A6",X"A8",X"AB",X"AD",X"AF",X"B2",X"B4",X"B7",
56  X"B9",X"BC",X"BF",X"C1",X"C4",X"C7",X"CA",X"CD",
57  X"D0",X"D2",X"D5",X"D8",X"DB",X"DE",X"E1",X"E4",
58  X"E8",X"EB",X"EE",X"F1",X"F4",X"F7",X"FA",X"FD",
59  X"00",X"03",X"06",X"09",X"0C",X"0F",X"12",X"15",
60  X"18",X"1C",X"1F",X"22",X"25",X"28",X"2B",X"2E",
61  X"30",X"33",X"36",X"39",X"3C",X"3F",X"41",X"44",
62  X"47",X"49",X"4C",X"4E",X"51",X"53",X"55",X"58",
63  X"5A",X"5C",X"5E",X"60",X"62",X"64",X"66",X"68",
64  X"6A",X"6C",X"6D",X"6F",X"70",X"72",X"73",X"75",
65  X"76",X"77",X"78",X"79",X"7A",X"7B",X"7C",X"7C",
66  X"7D",X"7E",X"7E",X"7F",X"7F",X"7F",X"7F",X"7F"
67 );
68
69 --pipeline registers
70 type pj_type is array(2 downto 0) of std_logic_vector(31 downto 0);
71 signal pj_reg : PJ_type;
72
73 type aj_type is array(4 downto 0) of std_logic_vector(7 downto 0);
74 signal aj_reg : AJ_type;
75
76 signal ena_reg : std_logic_vector(5 downto 0);
77
78 begin
79
80 process(CLK)
81   variable vxj : std_logic_vector(12 downto 0):=(others=>'0');
82   variable vxh : std_logic_vector(12 downto 0):=(others=>'0');
83   variable vyj : std_logic_vector(12 downto 0):=(others=>'0');
84   variable vyh : std_logic_vector(12 downto 0):=(others=>'0');
85   variable vdx : std_logic_vector(11 downto 0):=(others=>'0');
86   variable vdy : std_logic_vector(11 downto 0):=(others=>'0');
87   variable vdx2 : std_logic_vector(24 downto 0):=(others=>'0');
88   variable vdy2 : std_logic_vector(24 downto 0):=(others=>'0');
89   variable sign_expand : std_logic_vector(8 downto 0):=(others=>'0');
90 begin
91   if(R='1') then
92     SUBX_wire<=(others=>'0');
93     SUBY_wire<=(others=>'0');
94     SQRX_wire<=(others=>'0');
95     SQRY_wire<=(others=>'0');
96     ADDXY_wire<=(others=>'0');
97     MULTPJ_wire<=(others=>'0');
98     COS_wire<=(others=>'0');
99     MULTAJ_wire<=(others=>'0');
100    I<=(others=>'0');
101  elsif(CLK'event and CLK='1') then
102    vxj:=xj(11) & xj;
```

```
103        vxh:=xh(11) & xh;
104        vyj:=yj(11) & yj;
105        vyh:=yh(11) & yh;
106
107        SUBX_wire<=vxj-vxh;
108        SUBY_wire<=vyj-vyh;
109        if SUBX_wire(12)='1' then
110          vdx:=(not SUBX_wire(11 downto 0)) + 1;
111        else
112          vdx:=SUBX_wire(11 downto 0);
113        end if;
114        if SUBY_wire(12)='1' then
115          vdy:=(not SUBY_wire(11 downto 0)) + 1;
116        else
117          vdy:=SUBY_wire(11 downto 0);
118        end if;
119        SQRX_wire<=vdx*vdx;
120        SQRY_wire<=vdy*vdy;
121        vdx2:=SQRX_wire(23) & SQRX_wire;
122        vdy2:=SQRY_wire(23) & SQRY_wire;
123        ADDXY_wire<=vdx2+vdy2;
124        MULTPJ_wire<=ADDXY_wire*pj_reg(2);
125        COS_wire <= cos_tbl(conv_integer( MULTPJ_wire(31 downto 24) ));
126        MULTAJ_wire<=aj_reg(4)*COS_wire;
127        if ena_reg(5)='1' then
128          sign_expand := (others=>MULTAJ_wire(15));
129          I<=I+(sign_expand & MULTAJ_wire(15 downto 8));
130        end if;
131
132      end if;
133    end process;
134
135    -- pipeline registers
136    process (CLK, R)
137    begin
138      if(R = '1') then
139        for i in 0 to 4 loop
140          aj_reg(i) <= (others=>'0');
141        end loop;
142        for i in 0 to 2 loop
143          pj_reg(i) <= (others=>'0');
144        end loop;
145        ena_reg <= (others=>'0');
146      elsif (CLK'event and CLK='1') then
147        aj_reg(0) <= aj;
148        for i in 1 to 4 loop
149        aj_reg(i) <= aj_reg(i-1);
150        end loop;
```

```
151
152       pj_reg(0) <= pj;
153       for i in 1 to 2 loop
154         pj_reg(i) <= pj_reg(i-1);
155       end loop;
156
157       ena_reg(0) <= ena;
158       for i in 1 to 5 loop
159         ena_reg(i) <= ena_reg(i-1);
160       end loop;
161     end if;
162 end process;
163
164 end rtl;
```

A simulation waveform of Listing 6.4 is shown in Figure 6.8. The figures of Figure 6.8 are in hexadecimal notation. We calculated the intensity value I of the hologram at the hologram coordinates $(0xe00, 0xe80)(= (-512, -384))$ of Figure 6.5 using the pipeline. The notation $0x$ indicates that it is a hexadecimal notation. X_j, y_j, P_j, A_j when the control signal "ena" is at the high level and passes through the pipeline, whereby the intensity $I(0xe00, 0xe80)$ of the hologram is calculated. The calculation condition of this hologram is shown again. The size of the hologram is 1,024 × 768 pixels. The coordinates of the two object points are $(x_j, y_j) = (0xf00, 0x000)(= (-256, 0))$ and $(x_j, y_j) = (0x100, 0x000)(= (+256, 0))$. The wavelength is $\lambda = 633$ nm, the sampling pitch is p =10 μm, and the distance between the hologram and the object is 1 m. Under this condition, P_j is a 32-bit integer and becomes 0x00052D36. The calculation result is $I(0xe00, 0xe80) = 0x40(= 64)$, which matches the calculation result of the emulator.

6.1.3 PARALLEL IMPLEMENTATION

Each pixel on a hologram can be calculated in parallel by arranging the hologram calculation circuits. Figure 6.9 shows a block diagram of the parallel calculation circuit in which M pipeline circuits are implemented. By setting the hologram coordinates, $(x_h, y_h) \cdots (x_{h+M-1}, y_h)$, in registers, the intensity values of the hologram coordinates can be calculated in parallel.[v]

The object points data x_j, y_j, P_j, A_j are supplied from a shared memory to the pipeline. The shared memory is implemented in dynamic random access memories (DRAMs) or static RAMs (SRAMs) outside an FPGA or in internal memories inside an FPGA. Each block in Figure 6.9 operates with the same clock signal, and the information of the object points x_j, y_j, P_j, A_j for N points is read out by the address counter connected to the shared memory.

[v]The registers can be implemented using memory devices. They are usually implemented using D-Flip Flops.

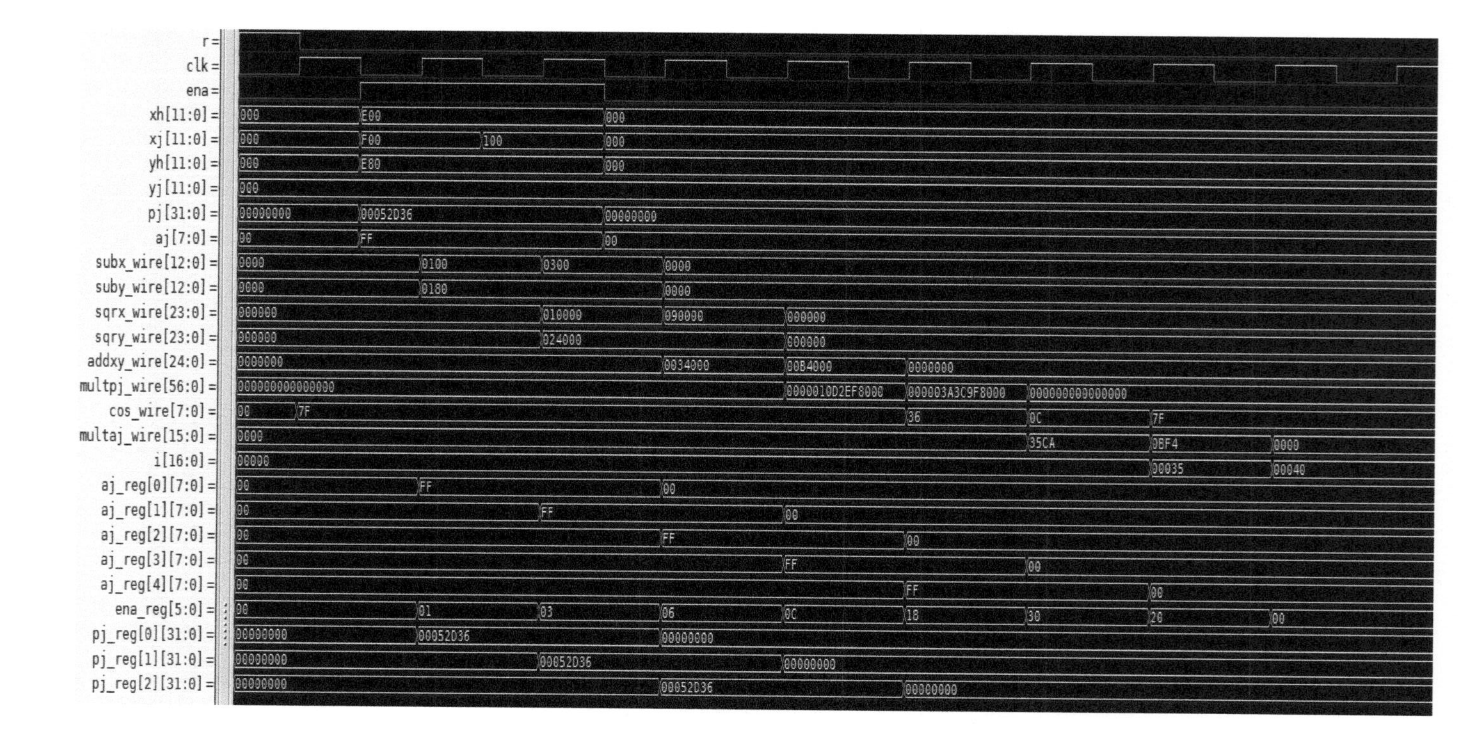

FIGURE 6.8 Simulation waveform of Listing 6.4.

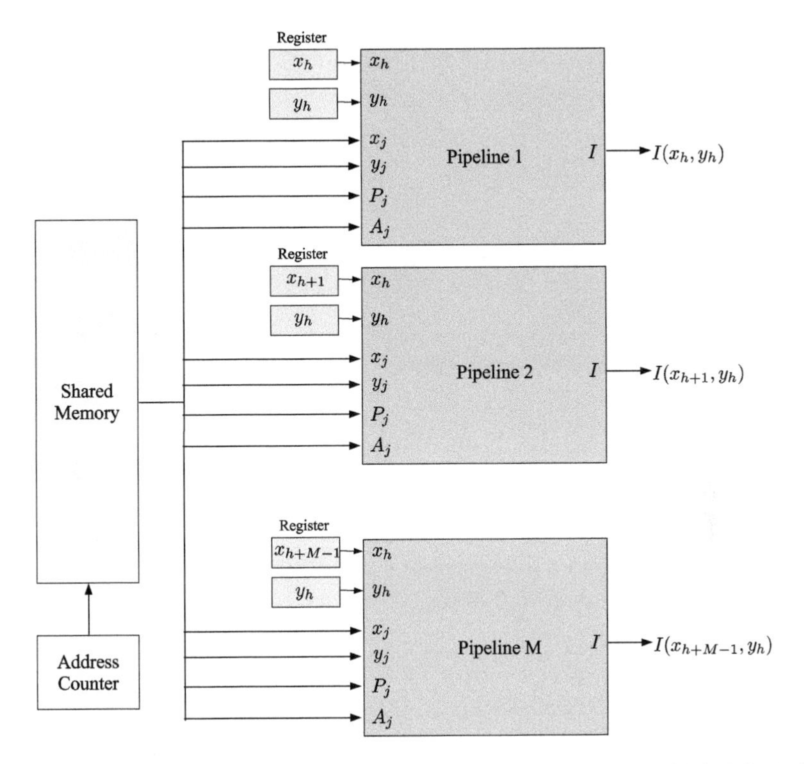

FIGURE 6.9 Block diagram of the parallel calculation circuit in which M pipeline circuits are implemented.

The operation of the address counter is started by the calculation start signal sent from the host computer. When the counter value reaches N, the address counter stops its operation and outputs a calculation end signal. When the host computer receives the calculation end signal, it reads the calculation results $I(x_h, y_h) \cdots I(x_{h+M-1}, y_h)$. For the communication interface between the host computer and the FPGA, we can use the PCI express bus or USB bus.

From HORN-4 onward, hologram calculation circuits are implemented more efficiently by using the recurrence formula introduced in Section 3.2.2. This recurrence formula can calculate the phase of light propagation only with adder circuits. In a hardware implementation, multiplying circuits have a large circuit area, so the circuit using the recurrence formula is advantageous. Here, we introduce the latest machine, HORN-8 [156]. In HORN-8, instead of a look-up table for trigonometric calculations, a method that can approximate trigonometric functions with simple bit operation is used [157].

Figure 6.10 is the HORN-8 board. Eight FPGAs are mounted on one board. One FPGA is for communication with the host computer, and the other seven FPGAs are for the calculation circuit. The FPGA is connected by a ring bus.

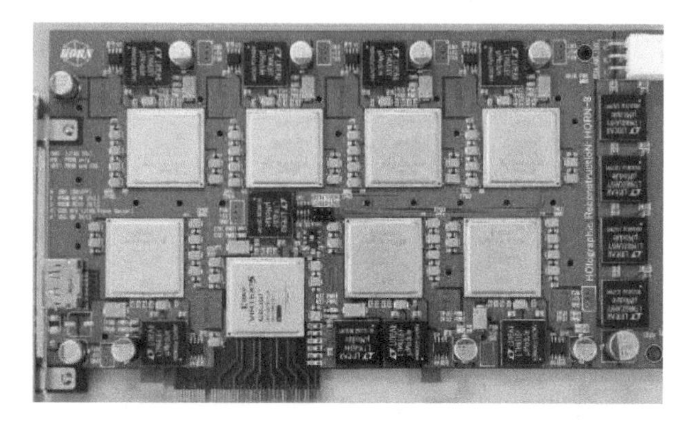

FIGURE 6.10 HORN-8 board.

TABLE 6.3
Calculation times when a hologram of 1,920×1,080 pixels is calculated from a 3D object composed of 10,000 object points.

System	Time/hologram	Speed ratio	Frame rate(fps)
CPU(Core i7-6700K, 4 cores, 4GHz)	2.951	1	0.34
GPU(GTX TITAN X, 3,072 cores, 1 GHz)	0.106	28	9.4
HORN-8 board(4,480 cores, 0.25 GHz)	0.019	155	53

In HORN-8, 640 hologram calculation circuits are implemented in the seven FPGAs. Therefore, one board has 4,480 hologram calculation circuits, and these circuits operate in parallel so that holograms can be calculated at high speed. The FPGA operates at 250 MHz. The host computer and the communication FPGA communicate via the PCI Express bus, and the transfer speed is 1.2 Gbytes/s. The communication time is hidden by overlapping the communication and hologram calculation.

Table 6.3 shows the calculation times when a hologram of 1,920 × 1,080 pixels is calculated from a 3D object composed of 10,000 object points. The hardware components we used are CPU, GPU (GPU is described in next chapter), and the HORN-8 board. The HORN-8 board is 160 times faster than the CPU. The HORN-8 board is 6 times faster than the GPU. The calculation FPGAs are Xilinx Virtex 5-XC5VLX110. The communication FPGA is Xilinx Virtex5-XC5VLX30T. Although these FPGAs are four generations ago, the hologram calculation is sufficiently fast compared to the latest GPU. In other words, if we use latest FPGAs, the calculation speed can be further increased.

6.1.4 DIFFRACTION CALCULATION ON FPGA

An FPGA-based computer FFT-HORN accelerates the Fresnel diffraction calculation in digital holography [158]. The first FFT-HORN reconstructed 100 images from a 256 × 256 pixel-hologram in 266 ms. The second FFT-HORN reconstructed 100 images from a 1,024 × 1,024 pixel-hologram in 3.3 ms. This clustered FFT-HORN required just 0.77 s to reconstruct 1,024 images from a 1,024 × 1,024 pixel hologram [158]. Other FPGA-based computers for digital holography have been developed by [159]. This computer calculates the Fresnel diffraction and the phase of a diffracted field.

Phase-shifting digital holography [160] reconstructs only object light from inline holograms, rejecting unwanted extraneous lights; however, the inline holograms are sequentially captured, which is unsuitable for moving objects. Moving objects can be adequately captured by parallel phase-shifting digital holography [161], which records inline holograms by one-shot recording. An FPGA-based computer that combines parallel phase-shifting digital holography and FFT-HORN has also been developed [162]. The number of bits required for implementing diffraction calculations using fixed-point arithmetic has been investigated in Ref. [163].

6.2 GPU IN COMPUTER HOLOGRAPHY

Although FPGA-based approaches significantly improve the speed of hologram and diffraction calculations, their practical use is hampered by the long development period for designing an FPGA board, long times of HDL compilation and mapping to an FPGA, and a high level of technical knowledge of FPGA technology.

Conversely, recently, **GPU**s with many simple processors that perform arithmetic operations and that are eminently suitable for parallel processing have been used for high-performance computing. The acceleration of numerical simulations by GPUs is referred to as general-purpose computation on GPU (**GPGPU**) or **GPU computing**. The merits of GPGPU include high computational power, and short compilation and development times. Figure 6.11 shows a modern GPU structure that consists of multiprocessors. The multi processor is further composed of "cores," " cache/shared memory" and "special function units." The cores perform 32- or 64-bit floating-point addition, multiplication, and multiply-add operations. "Cache/shared memory" is on-chip memory and is much faster than off-chip memory. It is used as a cache memory or a controllable fast small memory. A "special function units" is dedicated hardware that calculates frequently used mathematical functions (e.g., trigonometric functions, logarithmic functions, etc.) at high speed.

To the best of our knowledge, the first use of a graphics processing machine (graphics workstation) in a hologram calculation was reported by Ref. [164]. Since then, NVIDIA Geforce 4, Quadro and GeForce6 have been adopted in hologram calculations [165–167].

In 2007, NVIDIA released a programming environment called **Compute Unified Device Architecture** (**CUDA**) version 1.0. Previous GPUs that

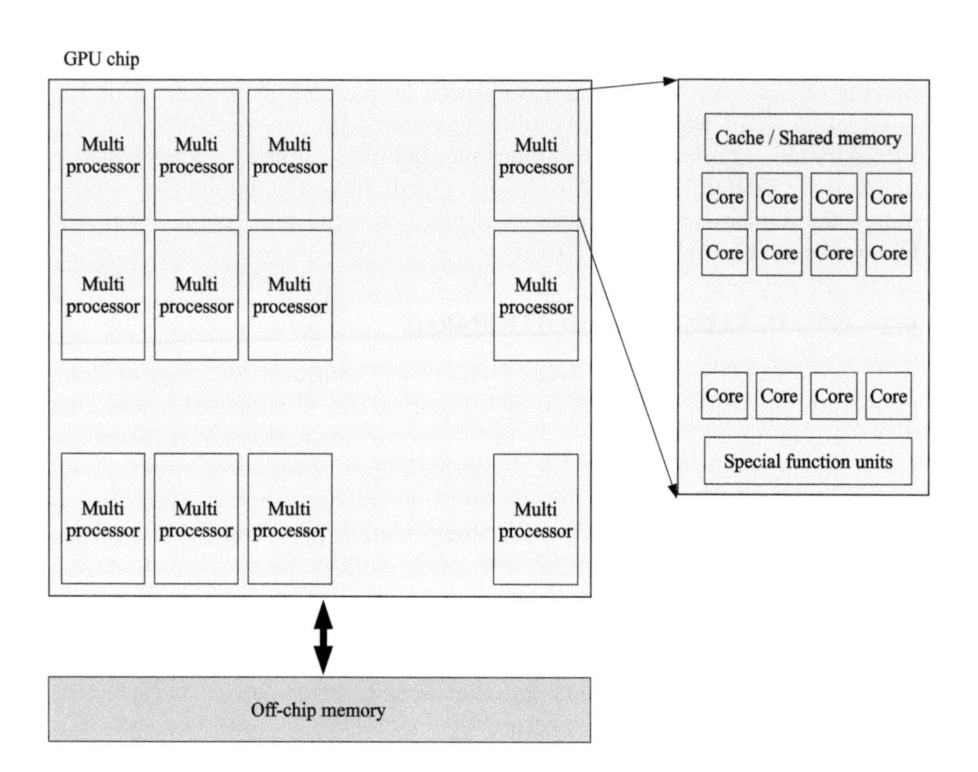

FIGURE 6.11 Pipeline circuit for hologram calculation.

do not support CUDA require graphics programming skills[vi] and GPU programming skills, which hindered the ability of users to take advantage of the high performance of GPUs.

Recent GPUs that support CUDA do not require graphics programming skills. A host computer controls the GPUs and the communication between the host computer and a GPU board via the PCI express bus and directly accesses the off-chip memory on the GPU board. The off-chip memory stores the input data and the results computed by a GPU chip. The instruction set of the GPU chip, called the kernel, is written in a C-type language source code and is compiled by a CUDA compiler. The kernel is downloaded to the GPU chip via the PCI express bus.

Most programs using CUDA calculate in the following steps.

- Allocate GPU memory.
- Transfer data from CPU to GPU.
- Start kernel function.
- Transfer calculation result from the GPU to CPU.
- Release the GPU memory.

Listing 6.5 shows a code simply adding 1 to an array of N elements using CUDA. The function simple_kernel is a function executed on a GPU. This function adds 1 to each element of GPU memory in parallel. A function executed by the GPU is called a kernel function. The function cudaMalloc reserves the GPU memory. The function cudaMemcpy transfers data from the CPU to the GPU or transfers data from the GPU to the CPU. The function cudaFree releases the GPU memory. In lines 15-17, we set the number of parallel executions of the kernel function, and the kernel function is executed. In this list, there is only one definition of the kernel function, but on the GPU a number of the kernel functions are executed on each core at the same time. The conceptual diagram of the parallel kernel execution is shown in Figure 6.12.

Listing 6.5 Simple CUDA code.

```
1  __global__ void simple_kernel(float *p){
2    int i = blockDim.x * blockIdx.x + threadIdx.x;
3    p[i] += 1;
4  }
5  void main(){
6    float *p_h, p_g;
7
8    p_h = (float *)malloc(sizeof(float) * N);
9    //prepare data in CPU memory
10   for(int i = 0; i < N; i++) p_h[i] = i;
11
```

[vi]For GPUs before CUDA, knowledge of a 3D graphics library was essential even for general-purpose calculation.

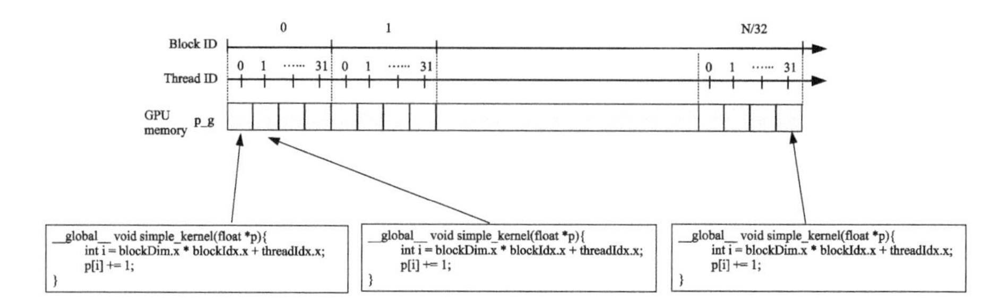

FIGURE 6.12 Conceptual diagram of the parallel kernel execution.

```
12    cudaMalloc((void**)&p_g, sizeof(float)*N);// allocate GPU memory
13    //transfer data from CPU memory to GPU memory
14    cudaMemcpy(p_g, p_h, sizeof(float)*N, cudaMemcpyHostToDevice);
15    //setup for the number of threads and blocks
16    dim3 blocks(N/32,1);
17    dim3 threads(32,1);
18    //execute kernel function
19    simple_kernel<<< blocks, threads >>>(p_g);
20    //transfer data from GPU memory to CPU memory
21    cudaMemcpy(p_h, p_g, sizeof(float)*N, cudaMemcpyDeviceToHost);
22    // free CPU memory
23    free(p_h);
24    // free GPU memory
25    cudaFree(gpu);
26  }
```

The latest GPUs possess thousands of cores and over 10 GB of device memory; thus, large-scale computer holography calculations can be performed by a single GPU board. However, a GPU requires a CPU as a host, and is more expensive than the latter, especially if multiple GPUs are involved.

As described in the previous sections, FFT is an essential algorithm in computer holography. CUDA provides a well-tuned FFT library called CUFFT, which is similar to the FFTW library[vii] and enables FFT calculations that are ten times faster than those on CPUs. In addition, CUDA provides a random number generator "cuRAND" for generating random phases, a linear algebra library "cuBLAS," and a C++ template class library "Thrust" for parallel reduction operations (such as searching for maximum and minimum values, and calculating sums and averages in parallel) and parallel sort algorithms.

Real-time DHM was probably the first application of CUDA to computer holography [168]. Since then, CUDA has been implemented in digital holography (Refs. [169,170]), CGH calculations (Ref. [171]), and holographic tweezers (Ref. [172]).

[vii]http://www.fftw.org/

6.2.1 DIFFRACTION CALCULATION USING GPU

In this section, let us describe how to implement diffraction calculation using a GPU. Fresnel diffraction described in Section 2.3.1 is rewritten as follows:

$$\begin{aligned} u_2(x_2, y_2) &= \frac{\exp(i\frac{2\pi}{\lambda}z)}{i\lambda z} \iint_{-\infty}^{+\infty} u_1(x_1, y_1) \times \\ &\quad \exp(i\frac{\pi}{\lambda z}((x_2 - x_1)^2 + (y_2 - y_1)^2))dx_1 dy_1, \\ &= \frac{\exp(i\frac{2\pi}{\lambda}z)}{i\lambda z} \times u_1(x_1, y_1) \otimes \exp(i\frac{\pi}{\lambda z}(x_1^2 + y_1^2)), \quad (6.3) \\ &= \frac{\exp(i\frac{2\pi}{\lambda}z)}{i\lambda z} \mathcal{F}^{-1}\left[\mathcal{F}\left[u_1(x_1, y_1)\right] \cdot \mathcal{F}\left[h_f(x_1, y_1)\right]\right], \quad (6.4) \end{aligned}$$

where $h_f(x, y)$ is the **impulse response** defined as

$$h_f(x, y) = \exp(i\frac{\pi}{\lambda z}(x^2 + y^2)). \quad (6.5)$$

Fresnel diffraction can be calculated by FFTs; therefore, in a GPU implementation, we use CUFFT library.

Listing 6.6 shows an entire source code of GPU-based Fresnel diffraction, ported from the CPU version of Fresnel diffraction (Listing 2.7). In this implementation, we ignore the complex coefficient of $\exp(i\frac{\pi}{\lambda z})$ in Eq. (6.4). The function "g_fresnel_fft" is a top function of the Fresnel diffraction. The arguments "u1" and "u2" are pointers of the source and destination planes. The arguments "Nx" and "Ny" are the size of the planes, and "lambda," "z," and "p" are the wavelength, propagation distance, and sampling pitch, respectively.

For simplicity, GPU memory management functions "cudaMalloc" and "cudaFree" are included in "g_fresnel_fft," but these memory functions are slow. If we further want to speed up "g_fresnel_fft," the memory functions should be executed in advance outside of "g_fresnel_fft."

Listing 6.6 Converting a double floating-point number to a fixed-point number.

```
1  __global__ void cu_fft_shift(cufftComplex *a, cufftComplex *b, int Nx,
       int Ny)
2  {
3    int x = blockIdx.x*blockDim.x + threadIdx.x;
4    int y = blockIdx.y*blockDim.y + threadIdx.y;
5
6    int NX_h = Nx / 2;
7    int NY_h = Ny / 2;
8    cufftComplex tmp1, tmp2;
9    int adr1, adr2;
10
11   adr1 = x + y*Nx;
```

```
12   adr2 = (x + NX_h) + (y + NY_h)*Nx;
13   tmp1 = a[adr1];
14   tmp2 = a[adr2];
15
16   b[adr1] = tmp2;
17   b[adr2] = tmp1;
18
19   adr1 = (x + NX_h) + y*Nx;
20   adr2 = x + (y + NY_h)*Nx;
21
22   tmp1 = a[adr1];
23   tmp2 = a[adr2];
24
25   b[adr1] = tmp2;
26   b[adr2] = tmp1;
27 }
28
29 void g_fft_shift(cufftComplex *a, cufftComplex *b, int Nx, int Ny)
30 {
31   dim3 block(32, 32, 1);
32   dim3 grid(Nx / 2 / block.x, Ny / 2 / block.y, 1);
33   cu_fft_shift <<< grid, block >>>(a, b, Nx, Ny);
34 }
35
36 __global__ void cu_response(cufftComplex *a, int Nx, int Ny, double
       param, double p)
37 {
38   int x = blockIdx.x*blockDim.x + threadIdx.x;
39   int y = blockIdx.y*blockDim.y + threadIdx.y;
40   int adr = (x + y*Nx);
41
42   float dx, dy;
43   float re, im;
44   dx = (float)(x - Nx / 2) * p;
45   dy = (float)(y - Ny / 2) * p;
46   float ph = param * (dx*dx + dy*dy);
47
48   re = __cosf(ph);
49   im = __sinf(ph);
50   a[adr] = make_cuComplex(re, im);
51 }
52
53 void g_response(cufftComplex *a, int Nx, int Ny, double lambda,
       double z, double p)
54 {
55   dim3 block(32, 32, 1);
56   dim3 grid(Nx / block.x, Ny / block.y, 1);
57   cu_response <<< grid, block >> >(a, Nx, Ny, M_PI/(lambda*z), p);
```

```
58 }
59
60 __global__ void cu_mul(cufftComplex *a, cufftComplex *b,
       cufftComplex *c, int Nx, int Ny)
61 {
62   int x = blockIdx.x*blockDim.x + threadIdx.x;
63   int y = blockIdx.y*blockDim.y + threadIdx.y;
64   int adr = (x + y*Nx);
65   float re, im;
66   re = cuCrealf(a[adr])*cuCrealf(b[adr]) - cuCimagf(a[adr])*cuCimagf(b[
         adr]);
67   im = cuCrealf(a[adr])*cuCimagf(b[adr]) + cuCimagf(a[adr])*cuCrealf(b[
         adr]);
68   c[adr] = make_cuComplex(re, im);
69 }
70
71 void g_mul_complex(cufftComplex *a, cufftComplex *b, cufftComplex
       *c, int Nx, int Ny)
72 {
73   dim3 block(32, 32, 1);
74   dim3 grid(Nx / block.x, Ny / block.y, 1);
75   cu_mul <<< grid, block >> >(a, b, c, Nx, Ny);
76 }
77 void g_fft(cufftComplex *src, cufftComplex *dst, int Nx, int Ny)
78 {
79   cufftHandle plan;
80   cufftPlan2d(&plan, Ny, Nx, CUFFT_C2C);
81   cufftExecC2C(plan, (cufftComplex*)src, (cufftComplex*)dst,
         CUFFT_FORWARD);
82   cufftDestroy(plan);
83 }
84
85
86 void g_ifft(cufftComplex *src, cufftComplex *dst, int Nx, int Ny)
87 {
88   cufftHandle plan;
89   cufftPlan2d(&plan, Ny, Nx, CUFFT_C2C);
90   cufftExecC2C(plan, (cufftComplex*)src, (cufftComplex*)dst,
         CUFFT_INVERSE);
91   cufftDestroy(plan);
92 }
93
94 void g_fresnel_fft(
95   cufftComplex* u1, cufftComplex* u2, int Nx, int Ny,
96   double lambda, double z, double p)
97 {
98   int size = sizeof(cufftComplex)*Nx*Ny;
99
```

```
100     cufftComplex *g_u1;
101     cufftComplex *g_u2;
102
103     cudaMalloc((void**)&g_u1, size);
104     cudaMalloc((void**)&g_u2, size);
105     cudaMemcpy(g_u1, u1, size, cudaMemcpyHostToDevice);
106
107     g_fft_shift(g_u1, g_u1, Nx, Ny);
108     g_fft(g_u1, g_u1, Nx, Ny);
109
110     g_response(g_u2, Nx, Ny, lambda, z, p);
111     g_fft(g_u2, g_u2, Nx, Ny);
112
113     g_mul_complex(g_u1, g_u2, g_u1, Nx, Ny);
114     g_ifft(g_u1, g_u1, Nx, Ny);
115     g_fft_shift(g_u1, g_u1, Nx, Ny);
116
117     cudaMemcpy(u2, g_u1, size, cudaMemcpyDeviceToHost);
118
119     cudaFree(g_u1);
120     cudaFree(g_u2);
121 }
```

OVERLAPPING THE COMMUNICATION BETWEEN CPU AND GPU USING STREAM

Listing 6.6 transfers the data from a CPU to a GPU, then performs diffraction calculation, and finally transfers the calculation result from the GPU to the CPU. If we want to continuously process new data with the GPU, the data transfer between the CPU and GPU may become a bottleneck. By using a technique called "stream," this data transfer can be hidden. A stream is a queue that manages kernel functions and data transfers. CUDA can create multiple streams, and if the GPU judges that the GPU can process each stream independently, each stream can run simultaneously on the GPU.

The timing chart when diffraction calculation is executed twice is shown in Figure 6.13. The upper part of the figure is when we do not use streams. The lower part is when using streams. Diffraction calculation in Figure 6.13 is performed by several kernel functions like Listing 6.6. If streams are not used, diffraction calculations are performed sequentially. On the other hand, when using streams, Data 2 can be transferred to the GPU while diffraction calculation of Data 1 is being executed. Further, while executing the diffraction calculation of Data 2, it is possible to transfer the diffraction calculation result of Data 1 to the CPU. In this way, by using the stream, the data transfer is concealed, and the total calculation time can be shortened.

The pseudo-code of diffraction calculation with/without using streams is shown in Figure 6.14. The left and right of the figure are pseudo-codes for

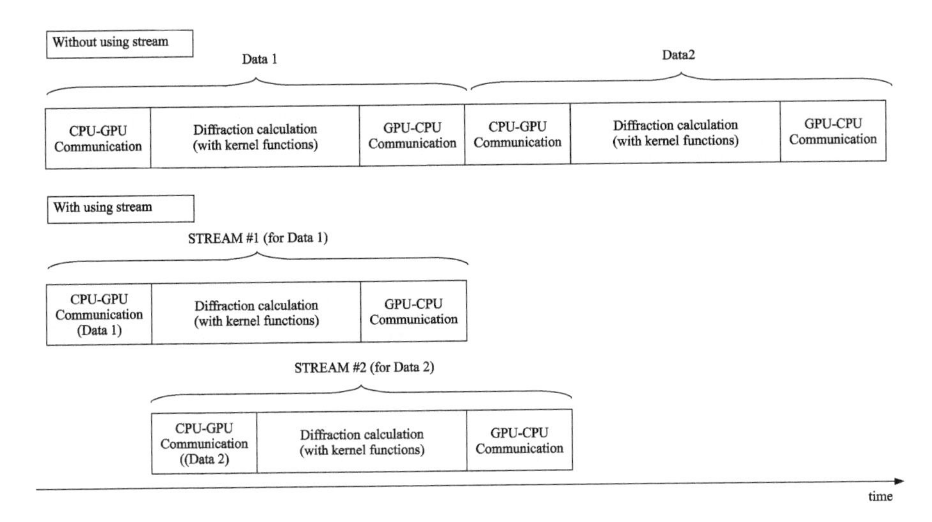

FIGURE 6.13 Timing chart when diffraction calculation is executed twice.

diffraction calculation when not using the stream and when using stream, respectively. If we do not use streams, we can use CUDA's "Sync API." On the other hand, if we are using streams, we will need to use the "Async API."

A timing chart when reconstructing images are actually calculated from holograms by diffraction calculation with three streams is shown in Figure 6.15.[viii] The hologram size is 512×512 pixels. If we do not use the stream (that is, we only use one stream), it takes 1.23 ms to reconstruct one hologram. Due to the effect of the stream, the communication time and diffraction calculation time can overlap successfully, and the reconstructing time of one hologram reduces 330 μs. As shown in the figure, it can be seen that a reconstructed image can be obtained at intervals of 330 μs in a pipeline manner.

6.3 COMPUTATIONAL WAVE OPTICS LIBRARY

Reference [173] reported a C++ class library (CWO++)[ix] for calculating wave optics in diffraction and hologram calculations that run on CPUs and GPUs. This library facilitates access to the computational performance of modern processors for optics engineers and researchers, who lack the technical knowledge of these processors. Here are some examples of using this library.

Listing 6.7 shows a sample code for using the CWO++ library with coding for diffraction calculations on a CPU with eight CPU threads. This source code performs diffraction calculation on the CPU, but by changing class "CWO"

[viii] We use a profiler that visualizes the internal operation of the GPU.

[ix] `http://cwolibrary.sourceforge.net/`

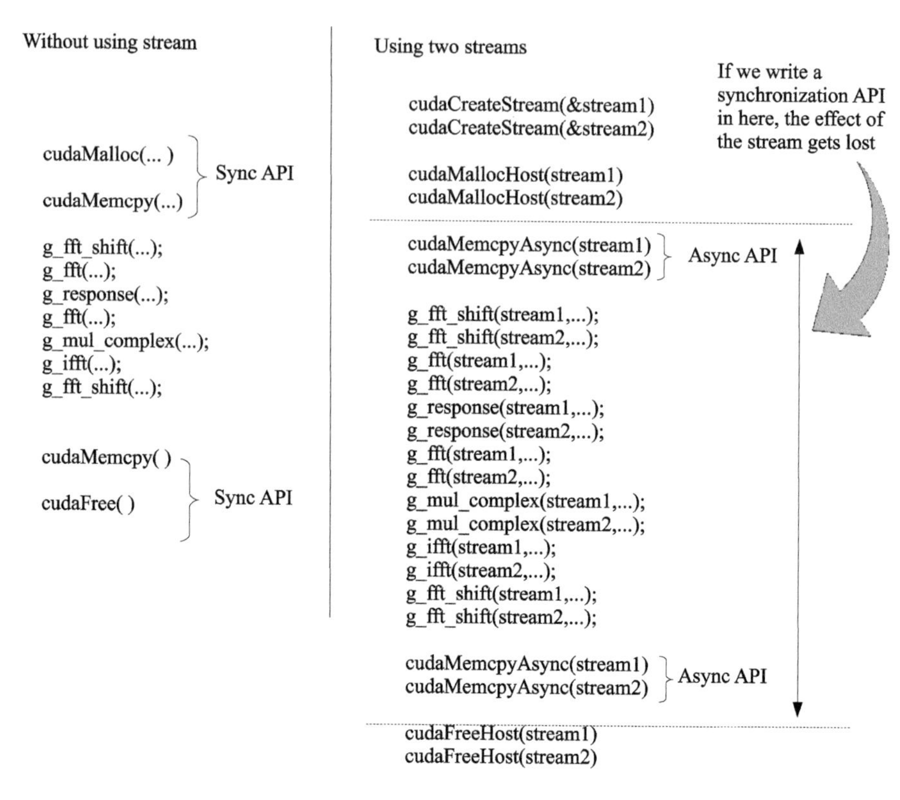

FIGURE 6.14 Pseudocodes of diffraction calculation with/without using streams.

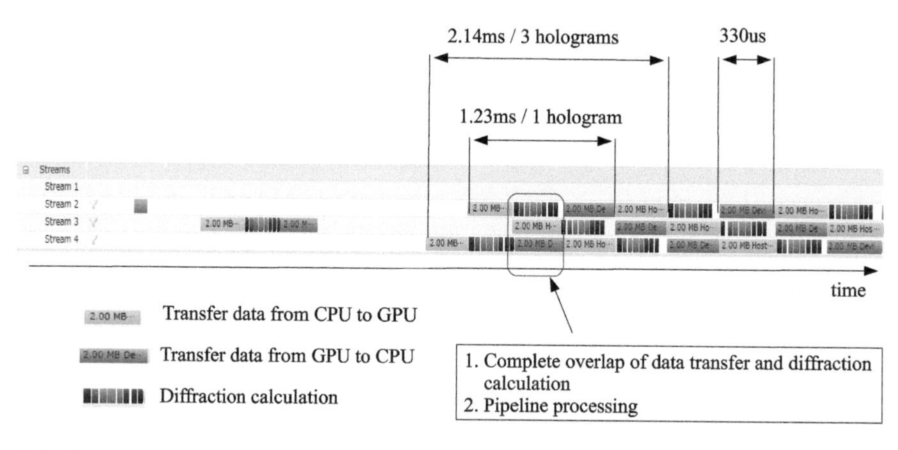

FIGURE 6.15 Timing chart when reconstructing images are actually calculated from holograms by diffraction calculation with three streams.

to class "GWO," this library is designed to execute all the calculations on the GPU. The function "Load" holds the image in "c." The function "SqrtReal" calculates the square root of the real part of the data currently held. The function "Diffract" performs a diffraction calculation. The first argument specifies the propagation distance, and the second argument specifies the type of diffraction calculation (in this case "CWO_ANGULAR" specifies the angle spectrum method described in Section 2.2).

Listing 6.7 Simple diffraction calculation using CWO++.

```
1 CWO c;
2 c.SetThreads(8); //Number of CPU threads (Default is 1)
3 c.Load("lena.bmp"); //1024x1024 pixels bitmap image
4 c.SqrtReal();//Taking square root
5 c.Diffract(0.1, CWO_ANGULAR); //angular spectrum method
6 //save diffracted intensity pattern as 256 gray scale bitmap file
7 c.SaveAsImage("diffruction_result_int.bmp",
      CWO_SAVE_AS_INTENSITY);
```

Listing 6.8 computes a kinoform and computes the reconstructed image from the kinoform. In kinoform, it is common to add a random phase to the original image. The random phase is added by the function "MulRandPhase."

Listing 6.8 Kinoform calculation using CWO++.

```
1 CWO c;
2 c.Load("lena.bmp"); //1024x1024 pixels bitmap image
3 c.SqrtReal();//Taking square root
4 c.MulRandPhase();
5 c.Diffract(0.1, CWO_ANGULAR); //angular spectrum method
6 c.Phase();//kinoform
7 c.SaveAsImage("kinoform.bmp",CWO_SAVE_AS_ARG);
8 c.Diffract(-0.1, CWO_ANGULAR); //reconstruction from the kinoform
9 c.Intensity();
10 //save diffracted intensity pattern as 256 gray scale bitmap file
11 c.SaveAsImage("reconstructed_image.bmp",CWO_SAVE_AS_RE);
```

Listing 6.9 is a program for 4-step digital holography described in Section 4.6. We calculate four inline holograms while changing the phase of the reference light by $\pi/4$ and calculate the object light from these holograms by Eq. (4.36). The reconstructed image obtained from this program is shown in Figure 4.22.

Listing 6.9 4-step phase-shifting digital holography using CWO++.

```
1 CWO c1,c2,c3,c4;
2 c1.Load("lena.bmp","mandrill.bmp");
3 c1.Diffract(0.3);
4 c1.ScaleCplx();
5 c2=c1;
```

```
6  c3=c1;
7  c4=c1;
8  //generating inline hologram 1
9  //add reference light with the phase shift of 0 radian
10 c1+=c1.Polar(1.0f, 0.0f);
11 c1.Intensity();
12 //generating inline hologram 2
13 //add reference light with the phase shift of pi/2
14 c2+=c2.Polar(1.0f, CWO_PI/2);
15 c2.Intensity();
16 //generating inline hologram 3
17 //add reference light with phase the shift of pi
18 c3+=c3.Polar(1.0f, CWO_PI);
19 c3.Intensity();
20 //generating inline hologram 4
21 //add reference light with the phase shift of 3pi/2
22 c4+=c4.Polar(1.0f, 3.0f/2.0f*CWO_PI);
23 c4.Intensity();
24 //retrieve object wave using 4-steps phase shifting digital
       holography
25 c1-=c3;
26 c2-=c4;
27 c3.Clear();
28 //The real and imaginary parts of c3 are replaced by
29 the real parts of c1 and c2, respectively
30 c3.ReIm(c1,c2);
31 //reconstruction
32 c3.Diffract(-0.3);
33 c3.SaveAsImage("phase_shift_lena.bmp",CWO_SAVE_AS_AMP);
34 c3.SaveAsImage("phase_shift_mandrill.bmp",CWO_SAVE_AS_ARG);
```

6.4 OTHER HARDWARE IN COMPUTER HOLOGRAPHY

In addition to FPGA and GPU architectures, computer holography has been implemented on a Xeon Phi (Figure 6.16) and Greatly Reduced Array of Processor Element with Data Reduction (GRAPE-DR) processors. Recently, Intel released the Xeon Phi processor that installs multiple x86-based processors on one chip. Reference [174] evaluated the performance of a Xeon Phi processor and compared diffraction and CGH calculations performed by Xeon Phi, CPU, and GPU. The GRAPE-DR processor is a multicore processor with 512 processor elements. The performance of GRAPE-DR in CGH calculations has been evaluated in Reference [175].

6.5 PROS AND CONS

Finally, we summarize the pros and cons of each hardware. Table 6.4 shows the number of cores of each hardware, clock frequency, power consumption,

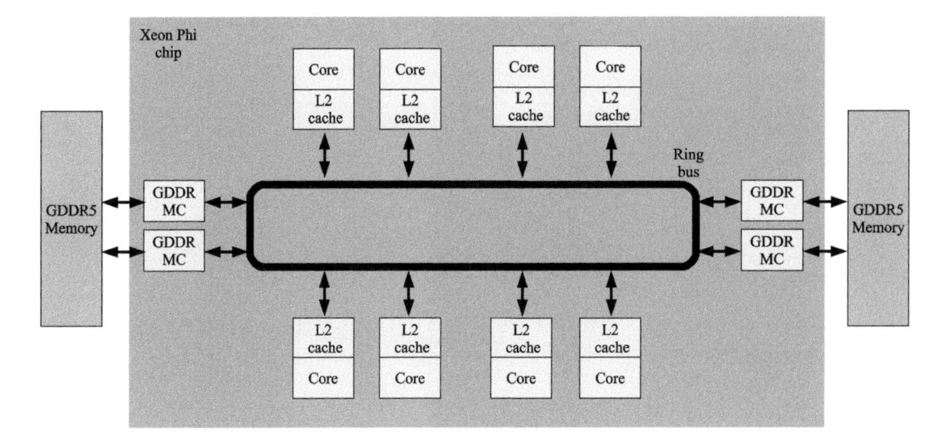

FIGURE 6.16 Xeon Phi coprocessor.

and development period.

MERITS AND DEMERITS OF CPU

The merits of CPUs are short development period, high clock frequency, easy programming, and flexibility. Double-precision floating-point arithmetic is useful for, e.g., an accurate distance calculation for CGH calculation. The drawbacks are a low parallelization that is proportional to the number of computational cores.

MERITS AND DEMERITS OF FPGA

The merits of FPGAs are the usage of pipeline architecture that enables calculations every clock cycle. Use of fixed-point format-based circuits, which have a small circuit area, leads to an effective practical use of hardware resources and high parallelization because FPGAs can implement massive numbers of computation cores for well-designed circuits. The demerits are the long development period for designing FPGA chips and boards, long times for HDL compilation and mapping to FPGA, and high required knowledge of FPGA technology.

MERITS AND DEMERITS OF GPU

Recent GPUs have over 2,000 computational cores, an average development period, and are easy to program compared with FPGAs. CUDA provides well-tuned libraries, for example, the FFT library CUFFT which is similar to the FFTW library and enables FFT calculations that are ten times faster than those on CPUs. The drawback is that the programming is somewhat more

TABLE 6.4

Pros and cons of each hardware.

	Number of cores	Development period	Clock frequency	Power consumption
CPU	Low	Short	Clock frequency	Average
FPGA	High	Long	Low	Low
GPU	High	Average	High	High
Xeon-phi	Average	Average	High	High

difficult compared with that for CPUs. A GPU requires a CPU as a host, and is more expensive than the latter, especially if multiple GPUs are involved.

MERITS AND DEMERITS OF XEON-PHI

The merit of the Xeon Phi is the capability to accelerate existing codes with some modifications. The drawbacks are a low parallelization compared with GPUs and FPGAs.

7 Appendix

In computer holography, the **Fourier transform** plays an important role. Here we summarize the formulas of the Fourier transform used in this textbook.

7.1 DEFINITION OF FOURIER TRANSFORM

Assuming that a one-dimensional signal is $u(x)$ and its Fourier transform is $U(f)$, the Fourier transform is defined as

$$U(f) = \int_{-\infty}^{\infty} u(x) \exp(-i2\pi f x) dx = \mathcal{F}\Big[u(x)\Big][u(x)]. \tag{7.1}$$

The **inverse Fourier transform** is defined as

$$u(x, y) = \int_{-\infty}^{\infty} U(f) \exp(i2\pi f x) df = \mathcal{F}^{-1}\Big[U(f)\Big]. \tag{7.2}$$

Some definitions require $1/2\pi$ or $1/\sqrt{2\pi}$ before the integration of the Fourier transform and inverse Fourier transform, but ignore them in this book.

Assuming that a two-dimensional signal is $u(x, y)$ and its Fourier transform is $U(f_x, f_y)$, the two-dimensional Fourier transform is defined as

$$U(f_x, f_y) = \int\int_{-\infty}^{\infty} u(x, y) \exp(-i2\pi(f_x x + f_y y)) dx dy = \mathcal{F}\Big[u(x, y)\Big]. \tag{7.3}$$

The inverse two-dimensional Fourier transform is defined as

$$u(x, y) = \int\int_{-\infty}^{\infty} U(f_x, f_y) \exp(i2\pi(f_x x + f_y y)) df_x df_y = \mathcal{F}^{-1}\Big[U(f_x, f_y)\Big]. \tag{7.4}$$

7.2 SHIFT THEOREM

The **shift theorem** of the Fourier transform is expressed as

$$u(x - x_0) \longleftrightarrow U(f) e^{-i2\pi f x_0} \tag{7.5}$$

$$u(x) \exp(i2\pi f_0 x) \longleftrightarrow U(f - f_0). \tag{7.6}$$

$\longrightarrow$ and $\longleftarrow$ denote the Fourier transform and the inverse Fourier transform, respectively.

$$u(x - x_0) \longleftrightarrow U(f) e^{-i2\pi f x_0} \tag{7.7}$$

$$u(x - x_0) \exp(i2\pi f_0 t) \longleftrightarrow U(f - f_0). \tag{7.8}$$

7.3 CONVOLUTION THEOREM

The **convolution theorem** is important because it is frequently used in diffraction calculations. The relationship between the Fourier transform and the convolution integral can be written as

$$u(x) * h(x) = \mathcal{F}^{-1}\left[\mathcal{F}\left[u(x)\right]\mathcal{F}\left[h(x)\right]\right]. \tag{7.9}$$

The proof is shown below:

$$\begin{aligned} F[u(x) * h(x)] &= \int_{-\infty}^{\infty}\left[\int_{-\infty}^{\infty} u(\xi)h(x-\xi)d\xi\right]e^{-i2\pi fx}dx \\ &= \int_{-\infty}^{\infty} u(\xi)\left[\int_{-\infty}^{\infty} h(x-\xi)e^{-i2\pi fx}dx\right]d\xi \\ &= \int_{-\infty}^{\infty} u(\xi)e^{-i2\pi f\xi}\mathcal{F}\left[h(x)\right]d\xi \quad \text{(Fourier shift theorem)} \\ &= \mathcal{F}\left[h(x)\right]\int_{-\infty}^{\infty} u(\xi)e^{-i2\pi f\xi}d\xi \\ &= \mathcal{F}\left[h(x)\right]\mathcal{F}\left[u(x)\right]. \end{aligned} \tag{7.10}$$

Therefore, by performing the inverse Fourier transformation on both sides, the convolution theorem is obtained as follows:

$$u(x) * h(x) = F^{-1}\left[F\left[u(x)\right]F\left[h(x)\right]\right]. \tag{7.11}$$

In a two-dimensional case, the convolution theorem is written as

$$u(x,y) * h(x,y) = \mathcal{F}^{-1}\left[\mathcal{F}\left[u(x,y)\right]\mathcal{F}\left[h(x,y)\right]\right]. \tag{7.12}$$

References

1. K. Matsushima and S. Nakahara. Extremely high-definition full-parallax computer-generated hologram created by the polygon-based method. *Appl. Opt.*, 48(34):H54–H63, 2009.
2. B. Kemper and G. Von Bally. Digital holographic microscopy for live cell applications and technical inspection. *Appl. Opt.*, 47(4):A52–A61, 2008.
3. T. Nakatsuji and K. Matsushima. Free-viewpoint images captured using phase-shifting synthetic aperture digital holography. *Appl. Opt.*, 47(19):D136–D143, 2008.
4. J. Garcia-Sucerquia. Color lensless digital holographic microscopy with micrometer resolution. *Opt. Lett.*, 37(10):1724–1726, May 2012.
5. I. Yamaguchi and T. Zhang. Phase-shifting digital holography. *Opt. Lett.*, 22(16):1268–1270, 1997.
6. Y. Awatsuji, M. Sasada, and T. Kubota. Parallel quasi-phase-shifting digital holography. *Appl. Phys. Lett.*, 85(6):1069–1071, 2004.
7. T. Kakue, R. Yonesaka, T. Tahara, Y. Awatsuji, K. Nishio, S. Ura, T. Kubota, and O. Matoba. High-speed phase imaging by parallel phase-shifting digital holography. *Opt. Lett.*, 36(21):4131–4133, Nov 2011.
8. P. Memmolo, C. Distante, M. Paturzo, A. Finizio, P. Ferraro, and B. Javidi. Automatic focusing in digital holography and its application to stretched holograms. *Opt. Lett.*, 36(10):1945–1947, 2011.
9. G. Pedrini, W. Osten, and Y. Zhang. Wave-front reconstruction from a sequence of interferograms recorded at different planes. *Opt. Lett.*, 30(8):833–835, 2005.
10. T. Shimobaba, M. Makowski, T. Kakue, M. Oikawa, N. Okada, Y. Endo, R. Hirayama, and T. Ito. Lensless zoomable holographic projection using scaled fresnel diffraction. *Opt. Express*, 21(21):25285–25290, Oct 2013.
11. T. Shimobaba, Y. Endo, R. Hirayama, Y. Nagahama, T. Takahashi, T. Nishitsuji, T. Kakue, A. Shiraki, N. Takada, N. Masuda, et al. Autoencoder-based holographic image restoration. *Appl. Opt.*, 56(13):F27–F30, 2017.
12. T. Shimobaba, N. Kuwata, M. Homma, T. Takahashi, Y. Nagahama, M. Sano, S. Hasegawa, R. Hirayama, T. Kakue, A. Shiraki, et al. Convolutional neural network-based data page classification for holographic memory. *Appl. Opt.*, 56(26):7327–7330, 2017.
13. J. W. Goodman. *Introduction to Fourier Optics.* Roberts and Company Publishers, 2005.
14. D. Gabor. A new microscopic principle. *Nature*, 161:777–778, 1948.
15. G. C. Sherman. Application of the convolution theorem to Rayleigh's integral formulas. *JOSA*, 57(4):546–547, 1967.
16. N. Delen and B. Hooker. Free-space beam propagation between arbitrarily oriented planes based on full diffraction theory: A fast Fourier transform approach. *JOSA A*, 15(4):857–867, Apr. 1998.
17. FFTW. `http://www.fftw.org/`.
18. E. Cuche, P. Marquet, and C. Depeursinge. Aperture apodization using cu-

bic spline interpolation: Application in digital holographic microscopy. *Opt. Commun.*, 182(1-3):59–69, 2000.
19. Y. Zhang, J. Zhao, Q. Fan, W. Zhang, and S. Yang. Improving the reconstruction quality with extension and apodization of the digital hologram. *Appl. Opt.*, 48(16):3070–3074, 2009.
20. S. Chang, D. Wang, Y. Wang, J. Zhao, and L. Rong. Improving the phase measurement by the apodization filter in the digital holography. In *Holography, Diffractive Optics, and Applications V*, volume 8556, page 85561S. International Society for Optics and Photonics, 2012.
21. S. Hasegawa, K. Shiono, and Y. Hayasaki. Femtosecond laser processing with a holographic line-shaped beam. *Opt. Express*, 23(18):23185–23194, 2015.
22. F. Dubois, O. Monnom, C. Yourassowsky, and J.-C. Legros. Border processing in digital holography by extension of the digital hologram and reduction of the higher spatial frequencies. *Appl. Opt.*, 41(14):2621–2626, 2002.
23. F. Zhang, I. Yamaguchi, and L. Yaroslavsky. Algorithm for reconstruction of digital holograms with adjustable magnification. *Opt. Lett.*, 29(14):1668–1670, Jul 2004.
24. R. P. Muffoletto, J. M. Tyler, and J. E. Tohline. Shifted Fresnel diffraction for computational holography. *Opt. Express*, 15(9):5631–5640, 2007.
25. J. F. Restrepo and J. Garcia-Sucerquia. Magnified reconstruction of digitally recorded holograms by Fresnel–Bluestein transform. *Appl. Opt.*, 49(33):6430–6435, Nov. 2010.
26. K. Matsushima. Shifted angular spectrum method for off-axis numerical propagation. *Opt. Express*, 18(17):18453–18463, Aug. 2010.
27. S. Odate, C. Koike, H. Toba, T. Koike, A. Sugaya, K. Sugisaki, K. Otaki, and K. Uchikawa. Angular spectrum calculations for arbitrary focal length with a scaled convolution. *Opt. Express*, 19(15):14268–14276, Jul. 2011.
28. T. Shimobaba, K. Matsushima, T. Kakue, N. Masuda, and T. Ito. Scaled angular spectrum method. *Opt. Lett.*, 37(19):4128–4130, 2012.
29. T. Shimobaba, T. Kakue, N. Okada, M. Oikawa, Y. Yamaguchi, and T. Ito. Aliasing-reduced Fresnel diffraction with scale and shift operations. *J. Opt.*, 15(7):075405, 2013.
30. T. Shimobaba, T. Kakue, M. Oikawa, N. Okada, Y. Endo, R. Hirayama, and T. Ito. Nonuniform sampled scalar diffraction calculation using nonuniform fast Fourier transform. *Opt. Lett.*, 38(23):5130–5133, Dec. 2013.
31. Y.-H. Kim, C.-W. Byun, H. Oh, J.-E. Pi, J.-H. Choi, G. H. Kim, M.-L. Lee, H. Ryu, and C.-S. Hwang. Off-axis angular spectrum method with variable sampling interval. *Opt. Commun.*, 348(0):31 – 37, Aug. 2015.
32. N. Okada, T. Shimobaba, Y. Ichihashi, R. Oi, K. Yamamoto, M. Oikawa, T. Kakue, N. Masuda, and T. Ito. Band-limited double-step Fresnel diffraction and its application to computer-generated holograms. *Opt. Express*, 21(7):9192–9197, 2013.
33. M. Stanley, M. A. Smith, A. P. Smith, P. J. Watson, S. D. Coomber, C. D. Cameron, C. W. Slinger, and A. Wood. 3d electronic holography display system using a 100-megapixel spatial light modulator. In *Optical Design and Engineering*, volume 5249, pages 297–309. International Society for Optics and Photonics, 2004.
34. C. Slinger, C. Cameron, and M. Stanley. Computer-generated holography as

a generic display technology. *Computer*, 38(8):46–53, 2005.

35. S. Tay, P.-A. Blanche, R. Voorakaranam, A. Tunc, W. Lin, S. Rokutanda, T. Gu, D. Flores, P. Wang, G. Li, et al. An updatable holographic three-dimensional display. *Nature*, 451(7179):694, 2008.
36. G. Nehmetallah and P. P. Banerjee. Applications of digital and analog holography in three-dimensional imaging. *Adv. Opt. Photonics*, 4(4):472–553, 2012.
37. M. E. Lucente. Interactive computation of holograms using a look-up table. *J. Electron. Imaging*, 2(1):28–34, Nov. 1993.
38. S.-C. Kim and E.-S. Kim. Effective generation of digital holograms of three-dimensional objects using a novel look-up table method. *Appl. Opt.*, 47(19):D55–D62, Jul 2008.
39. S.-C. Kim, J.-M. Kim, and E.-S. Kim. Effective memory reduction of the novel look-up table with one-dimensional sub-principle fringe patterns in computer-generated holograms. *Opt. Express*, 20(11):12021–12034, May. 2012.
40. M.-W. Kwon, S.-C. Kim, S.-E. Yoon, Y.-S. Ho, and E.-S. Kim. Object tracking mask-based NLUT on GPU for real-time generation of holographic videos of three-dimensional scenes. *Opt. Express*, 23(3):2101–2120, Feb. 2015.
41. Y. Pan, X. Xu, S. Solanki, X. Liang, R. B. A. Tanjung, C. Tan, and T.-C. Chong. Fast CGH computation using S-LUT on GPU. *Opt. Express*, 17(21):18543–18555, 2009.
42. J. Jia, Y. Wang, J. Liu, X. Li, Y. Pan, Z. Sun, B. Zhang, Q. Zhao, and W. Jiang. Reducing the memory usage for effective computer-generated hologram calculation using compressed look-up table in full-color holographic display. *Appl. Opt.*, 52(7):1404–1412, 2013.
43. K. Matsushima and M. Takai. Recurrence formulas for fast creation of synthetic three-dimensional holograms. *Appl. Opt.*, 39(35):6587–6594, Dec. 2000.
44. H. Yoshikawa. Fast computation of Fresnel holograms employing difference. *Opt. Rev.*, 8(5):331–335, May. 2001.
45. T. Shimobaba and T. Ito. An efficient computational method suitable for hardware of computer-generated hologram with phase computation by addition. *Comput. Phys. Commun.*, 138(1):44–52, Jul. 2001.
46. T. Shimobaba, S. Hishinuma, and T. Ito. Special-purpose computer for holography horn-4 with recurrence algorithm. *Comput. Phys. Commun.*, 148(2):160–170, Oct. 2002.
47. T. Yamaguchi, G. Okabe, and H. Yoshikawa. Real-time image plane full-color and full-parallax holographic video display system. *Opt. Eng.*, 46(12):125801–125801, Dec. 2007.
48. T. Yamaguchi and H. Yoshikawa. Computer-generated image hologram. *Chin. Opt. Lett.*, 9(12):120006, Dec. 2011.
49. T. Shimobaba, N. Masuda, and T. Ito. Simple and fast calculation algorithm for computer-generated hologram with wavefront recording plane. *Opt. Lett.*, 34(20):3133–3135, Oct. 2009.
50. T. Shimobaba, H. Nakayama, N. Masuda, and T. Ito. Rapid calculation algorithm of Fresnel computer-generated-hologram using look-up table and wavefront-recording plane methods for three-dimensional display. *Opt. Express*, 18(19):19504–19509, Sep. 2010.
51. P. Tsang, W.-K. Cheung, T.-C. Poon, and C. Zhou. Holographic video at 40 frames per second for 4-million object points. *Opt. Express*, 19(16):15205–

15211, Aug. 2011.

52. J. Weng, T. Shimobaba, N. Okada, H. Nakayama, M. Oikawa, N. Masuda, and T. Ito. Generation of real-time large computer generated hologram using wavefront recording method. *Opt. Express*, 20(4):4018–4023, Feb. 2012.
53. T. Shimobaba, T. Kakue, N. Masuda, Y. Ichihashi, K. Yamamoto, and T. Ito. Computer holography using wavefront recording method. *Digital Holography and Three-Dimensional Imaging (DH) DH2013, DTu1A*, 2, Apr. 2013.
54. A.-H. Phan, M. A. Alam, S.-H. Jeon, J.-H. Lee, and N. Kim. Fast hologram generation of long-depth object using multiple wavefront recording planes. In *Proc. SPIE*, volume 9006, pages 900612–900612. International Society for Optics and Photonics, Feb. 2014.
55. D. Arai, T. Shimobaba, K. Murano, Y. Endo, R. Hirayama, D. Hiyama, T. Kakue, and T. Ito. Acceleration of computer-generated holograms using tilted wavefront recording plane method. *Opt. Express*, 23(2):1740–1747, Jan. 2015.
56. P. Tsang and T.-C. Poon. Fast generation of digital holograms based on warping of the wavefront recording plane. *Opt. Express*, 23(6):7667–7673, Mar. 2015.
57. T. Shimobaba and T. Ito. Fast generation of computer-generated holograms using wavelet shrinkage. *Opt. Express*, 25(1):77–87, 2017.
58. D. Arai, T. Shimobaba, T. Nishitsuji, T. Kakue, N. Masuda, and T. Ito. An accelerated hologram calculation using the wavefront recording plane method and wavelet transform. *Opt. Commun.*, 393:107–112, 2017.
59. T. Shimobaba, K. Matsushima, T. Takahashi, Y. Nagahama, S. Hasegawa, M. Sano, R. Hirayama, T. Kakue, and T. Ito. Fast, large-scale hologram calculation in wavelet domain. *Opt. Commun.*, 412:80–84, 2018.
60. C. K. Chui. *An Introduction to Wavelets*. Elsevier, 2016.
61. D. L. Donoho. De-noising by soft-thresholding. *IEEE Transactions on Information Theory*, 41(3):613–627, 1995.
62. S. G. Mallat. A theory for multiresolution signal decomposition: The wavelet representation. *IEEE Transactions on Pattern Analysis and Machine Intelligence*, 11(7):674–693, 1989.
63. T. Tommasi and B. Bianco. Frequency analysis of light diffraction between rotated planes. *Opt. Lett.*, 17(8):556–558, Apr. 1992.
64. T. Tommasi and B. Bianco. Computer-generated holograms of tilted planes by a spatial frequency approach. *JOSA A*, 10(2):299–305, 1993.
65. K. Matsushima, H. Schimmel, and F. Wyrowski. Fast calculation method for optical diffraction on tilted planes by use of the angular spectrum of plane waves. *JOSA A*, 20(9):1755–1762, Sep. 2003.
66. K. Matsushima. Computer-generated holograms for three-dimensional surface objects with shade and texture. *Appl. Opt.*, 44(22):4607–4614, Aug. 2005.
67. K. Yamamoto, Y. Ichihashi, T. Senoh, R. Oi, and T. Kurita. Calculating the Fresnel diffraction of light from a shifted and tilted plane. *Opt. Express*, 20(12):12949–12958, Jun. 2012.
68. T. Yatagai. Stereoscopic approach to 3-d display using computer-generated holograms. *Appl. Opt.*, 15(11):2722–2729, Nov. 1976.
69. M. Yamaguchi, H. Hoshino, T. Honda, and N. Ohyama. Phase-added stereogram: calculation of hologram using computer graphics technique. In *Proc.*

SPIE, volume 1914, pages 25–31. International Society for Optics and Photonics, Sep. 1993.

70. H. Kang, T. Yamaguchi, and H. Yoshikawa. Accurate phase-added stereogram to improve the coherent stereogram. *Appl. Opt.*, 47(19):D44–D54, Jul. 2008.
71. Y. Takaki and K. Ikeda. Simplified calculation method for computer-generated holographic stereograms from multi-view images. *Opt. Express*, 21(8):9652–9663, Apr. 2013.
72. T. Mishina, M. Okui, and F. Okano. Calculation of holograms from elemental images captured by integral photography. *Appl. Opt.*, 45(17):4026–4036, Jun. 2006.
73. Y. Ichihashi, R. Oi, T. Senoh, K. Yamamoto, and T. Kurita. Real-time capture and reconstruction system with multiple GPUs for a 3d live scene by a generation from 4k IP images to 8k holograms. *Opt. Express*, 20(19):21645–21655, Sep. 2012.
74. J.-H. Park, K. Hong, and B. Lee. Recent progress in three-dimensional information processing based on integral imaging. *Appl. Opt.*, 48(34):H77–H94, 2009.
75. K. Wakunami and M. Yamaguchi. Calculation for computer generated hologram using ray-sampling plane. *Opt. Express*, 19(10):9086–9101, May. 2011.
76. J.-H. Park, M.-S. Kim, G. Baasantseren, and N. Kim. Fresnel and Fourier hologram generation using orthographic projection images. *Opt. Express*, 17(8):6320–6334, 2009.
77. S. Igarashi, T. Nakamura, K. Matsushima, and M. Yamaguchi. Efficient tiled calculation of over-10-gigapixel holograms using ray-wavefront conversion. *Opt. Express*, 26(8):10773–10786, 2018.
78. H. Zhang, Y. Zhao, L. Cao, and G. Jin. Fully computed holographic stereogram based algorithm for computer-generated holograms with accurate depth cues. *Opt. Express*, 23(4):3901–3913, 2015.
79. L. Xu, C. Chang, S. Feng, C. Yuan, J. Xia, J. You, and S. Nie. Stereoscopic hologram calculation based on nonuniform sampled wavefront recording plane. *Opt. Commun.*, 426:194–200, 2018.
80. J.-S. Chen, D. Chu, and Q. Y. Smithwick. Rapid hologram generation utilizing layer-based approach and graphic rendering for realistic three-dimensional image reconstruction by angular tiling. *J. Electronic Imaging*, 23(2):023016, 2014.
81. H. Zhang, Y. Zhao, L. Cao, and G. Jin. Layered holographic stereogram based on inverse Fresnel diffraction. *Appl. Opt.*, 55(3):A154–A159, 2016.
82. S.-K. Lee, S.-I. Hong, Y.-S. Kim, H.-G. Lim, N.-Y. Jo, and J.-H. Park. Hologram synthesis of three-dimensional real objects using portable integral imaging camera. *Opt. Express*, 21(20):23662–23670, 2013.
83. Y. Endo, K. Wakunami, T. Shimobaba, T. Kakue, D. Arai, Y. Ichihashi, K. Yamamoto, and T. Ito. Computer-generated hologram calculation for real scenes using a commercial portable plenoptic camera. *Opt. Commun.*, 356:468–471, 2015.
84. R. Oi, K. Yamamoto, and M. Okui. Electronic generation of holograms by using depth maps of real scenes. In *Proc. SPIE*, volume 6912, page 69120M. International Society for Optics and Photonics, Jan. 2008.
85. T. Shimobaba, Y. Nagahama, T. Kakue, N. Takada, N. Okada, Y. Endo,

R. Hirayama, D. Hiyama, and T. Ito. Calculation reduction method for color digital holography and computer-generated hologram using color space conversion. *Opt. Eng.*, 53(2):024108, 2014.

86. D. Hiyama, T. Shimobaba, T. Kakue, and T. Ito. Acceleration of color computer-generated hologram from RGB-D images using color space conversion. *Opt. Commun.*, 340:121–125, 2015.
87. T. Shimobaba, M. Makowski, Y. Nagahama, Y. Endo, R. Hirayama, D. Hiyama, S. Hasegawa, M. Sano, T. Kakue, M. Oikawa, et al. Color computer-generated hologram generation using the random phase-free method and color space conversion. *Appl. Opt.*, 55(15):4159–4165, 2016.
88. Y. Ogihara, T. Ichikawa, and Y. Sakamoto. Fast calculation with point-based method to make computer-generated holograms of the polygon model. In *Proc. SPIE*, volume 9006, page 90060T. International Society for Optics and Photonics, Feb. 2014.
89. Y. Sakamoto and M. Tobise. Computer generated cylindrical hologram. In *Proc. SPIE*, volume 57425, pages 267–274. International Society for Optics and Photonics, May. 2005.
90. A. Kashiwagi and Y. Sakamoto. A fast calculation method of cylindrical computer-generated holograms which perform image-reconstruction of volume data. In *Digital Holography and Three-Dimensional Imaging*, page DWB7. Optical Society of America, Jun. 2007.
91. T. Yamaguchi, T. Fujii, and H. Yoshikawa. Fast calculation method for computer-generated cylindrical holograms. *Appl. Opt.*, 47(19):D63–D70, Jul. 2008.
92. T. Yamaguchi, T. Fujii, and H. Yoshikawa. Computer-generated cylindrical rainbow hologram. *Opt. Eng.*, 48(5):055801, 2009.
93. B. J. Jackin and T. Yatagai. Fast calculation method for computer-generated cylindrical hologram based on wave propagation in spectral domain. *Opt. Express*, 18(25):25546–25555, Dec. 2010.
94. M. Oikawa, T. Shimobaba, N. Masuda, and T. Ito. Computer-generated hologram using an approximate Fresnel integral. *J. Opt.*, 13(7):075405, Jun. 2011.
95. G. B. Esmer. Performance assessment of a fast and accurate scalar optical diffraction field computation algorithm. *3D Research*, 4(1):1–7, Dec. 2013.
96. F. Yaraş, H. Kang, and L. Onural. State of the art in holographic displays: a survey. *J. Disp. Tech.*, 6(10):443–454, Oct. 2010.
97. P. Tsang, J. Liu, K. Cheung, and T.-C. Poon. Modern methods for fast generation of digital holograms. *3D Research*, 1(2):11–18, Nov. 2010.
98. J. Liu, J. Jia, Y. Pan, and Y. Wang. Overview of fast algorithm in 3d dynamic holographic display. In *Proc. SPIE*, volume 8913, page 89130X. International Society for Optics and Photonics, Aug. 2013.
99. T. Shimobaba, T. Kakue, and T. Ito. Review of fast algorithms and hardware implementations on computer holography. *IEEE Tran. Ind. Info.*, 12(4):1611–1622, 2016.
100. T. Nishitsuji, T. Shimobaba, T. Kakue, and T. Ito. Review of fast calculation techniques for computer-generated holograms with the point-light-source-based model. *IEEE Trans.Ind. Info.*, 13(5):2447–2454, 2017.
101. M. Takeda, H. Ina, and S. Kobayashi. Fourier-transform method of fringe-

pattern analysis for computer-based topography and interferometry. *JOSA*, 72(1):156–160, 1982.

102. E. Cuche, P. Marquet, and C. Depeursinge. Spatial filtering for zero-order and twin-image elimination in digital off-axis holography. *Appl. Opt.*, 39(23):4070–4075, Aug 2000.
103. T. Shimobaba, H. Yamanashi, T. Kakue, M. Oikawa, N. Okada, Y. Endo, R. Hirayama, N. Masuda, and T. Ito. In-line digital holographic microscopy using a consumer scanner. *Sci. Rep.*, 3, 2013.
104. I. Yamaguchi, J.-i. Kato, S. Ohta, and J. Mizuno. Image formation in phase-shifting digital holography and applications to microscopy. *Appl. Opt.*, 40(34):6177–6186, 2001.
105. C.-S. Guo, L. Zhang, H.-T. Wang, J. Liao, and Y. Zhu. Phase–shifting error and its elimination in phase-shifting digital holography. *Opt. Lett.*, 27(19):1687–1689, 2002.
106. L. Martinez-Leon, M. Araiza-E, B. Javidi, P. Andres, V. Climent, J. Lancis, and E. Tajahuerce. Single-shot digital holography by use of the fractional talbot effect. *Opt. Express*, 17(15):12900–12909, 2009.
107. A. Siemion, M. Sypek, M. Makowski, J. Suszek, A. Siemion, D. Wojnowski, and A. Kolodziejczyk. One-exposure phase-shifting digital holography based on the self-imaging effect. *Opt. Eng.*, 49(5):055802–055802–5, 2010.
108. M. Imbe and T. Nomura. Single-exposure phase-shifting digital holography using a random-complex-amplitude encoded reference wave. *Appl. Opt.*, 52(1):A161–A166, Jan 2013.
109. K. Maejima and K. Sato. One-shot digital holography for real-time recording of moving color 3-d images. *Advances in Imaging*, page DMA2, 2009.
110. Y. Han and Q. Yue. Laplacian differential reconstruction of one in-line digital hologram. *Opt. Commun.*, 283(6):929–931, 2010.
111. Z. Ren, N. Chen, A. Chan, and E. Y. Lam. Autofocusing of optical scanning holography based on entropy minimization. In *Digital Holography and Three-Dimensional Imaging*, pages DT4A–4. Optical Society of America, 2015.
112. S. Jiao, P. W. M. Tsang, T.-C. Poon, J.-P. Liu, W. Zou, and X. Li. Enhanced autofocusing in optical scanning holography based on hologram decomposition. *IEEE Transactions on Industrial Informatics*, 2017.
113. P. Langehanenberg, B. Kemper, D. Dirksen, and G. Von Bally. Autofocusing in digital holographic phase contrast microscopy on pure phase objects for live cell imaging. *Appl. Opt.*, 47(19):D176–D182, 2008.
114. M. Liebling and M. Unser. Autofocus for digital Fresnel holograms by use of a Fresnelet-sparsity criterion. *JOSA A*, 21(12):2424–2430, 2004.
115. J. R. Fienup. Phase retrieval algorithms: A comparison. *Appl. Opt.*, 21(15):2758–2769, 1982.
116. H. Kogelnik. Coupled wave theory for thick hologram gratings. *Bell System Technical Journal*, 48(9):2909–2947, 1969.
117. L. Hesselink, S. S. Orlov, and M. C. Bashaw. Holographic data storage systems. *Proceedings of the IEEE*, 92(8):1231–1280, 2004.
118. P. F. Van Kessel, L. J. Hornbeck, R. E. Meier, and M. R. Douglass. A mems-based projection display. *Proc. IEEE*, 86(8):1687–1704, 1998.
119. E. Buckley. Holographic laser projection. *J. Disp. Tech.*, 7(3):135–140, 2011.
120. T. Shimobaba, T. Kakue, and T. Ito. Real-time and low speckle holographic

projection. In *Industrial Informatics (INDIN), 2015 IEEE 13th International Conference on*, pages 732–741. IEEE, 2015.

121. J. Amako, H. Miura, and T. Sonehara. Speckle-noise reduction on kinoform reconstruction using a phase-only spatial light modulator. *Appl. Opt.*, 34(17):3165–3171, Jun. 1995.
122. A. W. Lohmann and D. Paris. Binary Fraunhofer holograms, generated by computer. *Appl. Opt.*, 6(10):1739–1748, 1967.
123. T. Shimobaba and T. Ito. Random phase-free computer-generated hologram. *Opt. Express*, 23(7):9549–9554, 2015.
124. T. Shimobaba, T. Kakue, Y. Endo, R. Hirayama, D. Hiyama, S. Hasegawa, Y. Nagahama, M. Sano, M. Oikawa, T. Sugie, et al. Random phase-free kinoform for large objects. *Opt. Express*, 23(13):17269–17274, 2015.
125. T. Shimobaba, T. Kakue, Y. Endo, R. Hirayama, D. Hiyama, S. Hasegawa, Y. Nagahama, M. Sano, M. Oikawa, T. Sugie, et al. Improvement of the image quality of random phase-free holography using an iterative method. *Opt. Commun.*, 355:596–601, 2015.
126. T. Shimobaba, T. Kakue, and T. Ito. Random phase-free computer holography and its applications. In *Three-Dimensional Imaging, Visualization, and Display 2016*, volume 9867, page 98670M. International Society for Optics and Photonics, 2016.
127. H. J. Coufal, D. Psaltis, G. T. Sincerbox, et al. *Holographic Data Storage*, volume 8. Springer, 2000.
128. P. Refregier and B. Javidi. Optical image encryption based on input plane and Fourier plane random encoding. *Opt. Lett.*, 20(7):767–769, 1995.
129. G. Situ and J. Zhang. Double random-phase encoding in the Fresnel domain. *Opt. Lett.*, 29(14):1584–1586, 2004.
130. J. F. Barrera, A. Mira, and R. Torroba. Optical encryption and QR codes: secure and noise-free information retrieval. *Opt. Express*, 21(5):5373–5378, 2013.
131. J. Shuming and P. Tsang. An iterative algorithm for holographic-QR (H-QR) code damage restoration. In *2015 IEEE 13th International Conference on Industrial Informatics (INDIN)*, pages 682–685. IEEE, 2015.
132. S. Jiao, W. Zou, and X. Li. QR code based noise-free optical encryption and decryption of a gray scale image. *Opt. Commun.*, 387:235–240, 2017.
133. G. E. Hinton and R. R. Salakhutdinov. Reducing the dimensionality of data with neural networks. *Science*, 313(5786):504–507, 2006.
134. P. Vincent, H. Larochelle, Y. Bengio, and P.-A. Manzagol. Extracting and composing robust features with denoising autoencoders. In *Proceedings of the 25th International Conference on Machine Learning*, pages 1096–1103. ACM, 2008.
135. D. Kingma and J. Ba. Adam: A method for stochastic optimization. *arXiv preprint arXiv:1412.6980*, 2014.
136. N. Srivastava, G. E. Hinton, A. Krizhevsky, I. Sutskever, and R. Salakhutdinov. Dropout: A simple way to prevent neural networks from overfitting. *Journal of Machine Learning Research*, 15(1):1929–1958, 2014.
137. S. Tokui, K. Oono, S. Hido, and J. Clayton. Chainer: A next-generation open source framework for deep learning. In *Proceedings of Workshop on Machine Learning Systems (LearningSys) in The Twenty-Ninth Annual Conference on*

Neural Information Processing Systems (NIPS), 2015.
138. G. W. Burr and T. Weiss. Compensation for pixel misregistration in volume holographic data storage. *Opt. Lett.*, 26(8):542–544, 2001.
139. H. Ruan. Recent advances in holographic data storage. *Frontiers of Optoelectronics*, 7(4):450–466, 2014.
140. J. F. Heanue, K. Gürkan, and L. Hesselink. Signal detection for page-access optical memories with intersymbol interference. *Appl. Opt.*, 35(14):2431–2438, 1996.
141. S.-H. Lee, S.-Y. Lim, N. Kim, N.-C. Park, H. Yang, K.-S. Park, and Y.-P. Park. Increasing the storage density of a page-based holographic data storage system by image upscaling using the PSF of the Nyquist aperture. *Opt. Express*, 19(13):12053–12065, 2011.
142. D.-H. Kim, S. Jeon, N.-C. Park, and K.-S. Park. Iterative design method for an image filter to improve the bit error rate in holographic data storage systems. *Microsystem Technologies*, 20(8-9):1661–1669, 2014.
143. G. W. Burr, J. Ashley, H. Coufal, R. K. Grygier, J. A. Hoffnagle, C. M. Jefferson, and B. Marcus. Modulation coding for pixel-matched holographic data storage. *Opt. Lett.*, 22(9):639–641, 1997.
144. A. Krizhevsky, I. Sutskever, and G. E. Hinton. Imagenet classification with deep convolutional neural networks. In *Advances in Neural Information Processing Systems*, pages 1097–1105, 2012.
145. T.-C. Poon. *Digital Holography and Three–Dimensional Display: Principles and Applications*. Springer Science & Business Media, 2006.
146. T. Shimobaba, T. Kakue, and T. Ito. Convolutional neural network-based regression for depth prediction in digital holography. *arXiv preprint arXiv:1802.00664*, 2018.
147. T. Pitkäaho, A. Manninen, and T. J. Naughton. Performance of autofocus capability of deep convolutional neural networks in digital holographic microscopy. In *Digital Holography and Three-Dimensional Imaging*, pages W2A–5. Optical Society of America, 2017.
148. T. Pitkäaho, A. Manninen, and T. J. Naughton. Focus classification in digital holographic microscopy using deep convolutional neural networks. In *European Conference on Biomedical Optics*, page 104140K. Optical Society of America, 2017.
149. T. Ito, T. Yabe, M. Okazaki, and M. Yanagi. Special-purpose computer horn-1 for reconstruction of virtual image in three dimensions. *Comput. Phys. Commun.*, 82(2):104–110, Sep. 1994.
150. T. Ito, H. Eldeib, K. Yoshida, S. Takahashi, T. Yabe, and T. Kunugi. Special-purpose computer for holography horn-2. *Comput. Phys. Commun.*, 93(1):13–20, Jan. 1996.
151. T. Shimobaba, N. Masuda, T. Sugie, S. Hosono, S. Tsukui, and T. Ito. Special-purpose computer for holography horn-3 with pld technology. *Comput. Phys. Commun.*, 130(1):75–82, Jul. 2000.
152. T. Ito, N. Masuda, K. Yoshimura, A. Shiraki, T. Shimobaba, and T. Sugie. Special-purpose computer horn-5 for a real-time electroholography. *Opt. Express*, 13(6):1923–1932, Mar. 2005.
153. Y. Ichihashi, H. Nakayama, T. Ito, N. Masuda, T. Shimobaba, A. Shiraki, and T. Sugie. Horn-6 special-purpose clustered computing system for elec-

troholography. *Opt. Express*, 17(16):13895–13903, Aug. 2009.

154. Y.-H. Seo, H.-J. Choi, J.-S. Yoo, and D.-W. Kim. Cell-based hardware architecture for full-parallel generation algorithm of digital holograms. *Opt. Express*, 19(9):8750–8761, Apr. 2011.
155. Y.-H. Seo, Y.-H. Lee, J.-S. Yoo, and D.-W. Kim. Hardware architecture of high-performance digital hologram generator on the basis of a pixel-by-pixel calculation scheme. *Appl. Opt.*, 51(18):4003–4012, Jun. 2012.
156. T. Sugie, T. Akamatsu, T. Nishitsuji, R. Hirayama, N. Masuda, H. Nakayama, Y. Ichihashi, A. Shiraki, M. Oikawa, N. Takada, Y. Endo, T. Kakue, T. Shimobaba, and T. Ito. High-performance parallel computing for next-generation holographic imaging. *Nature Electronics*, 1:254–259, 2018.
157. T. Nishitsuji, T. Shimobaba, T. Kakue, D. Arai, and T. Ito. Simple and fast cosine approximation method for computer-generated hologram calculation. *Opt. Express*, 23(25):32465–32470, 2015.
158. N. Masuda, T. Sugie, T. Ito, S. Tanaka, Y. Hamada, S.-i. Satake, T. Kunugi, and K. Sato. Special purpose computer system with highly parallel pipelines for flow visualization using holography technology. *Comput. Phys. Commun.*, 181(12):1986–1989, Dec. 2010.
159. C.-J. Cheng, W.-J. Hwang, C.-T. Chen, and X.-J. Lai. Efficient FPGA-based Fresnel transform architecture for digital holography. *J. Disp. Tech.*, 10(4):272–281, Apr. 2014.
160. I. Yamaguchi and T. Zhang. Phase-shifting digital holography. *Opt. Lett.*, 22(16):1268–1270, Aug. 1997.
161. Y. Awatsuji, M. Sasada, and T. Kubota. Parallel quasi-phase-shifting digital holography. *Appl. Phys. Lett.*, 85(6):1069–1071, Aug. 2004.
162. T. Kakue, T. Shimobaba, and T. Ito. High-speed parallel phase-shifting digital holography system using special-purpose computer for image reconstruction. In *SPIE Sensing Technology + Applications*, pages 9495–22, Apr. 2015.
163. N. Pandey and B. Hennelly. Fixed-point numerical-reconstruction for digital holographic microscopy. *Opt. Lett.*, 35(7):1076–1078, Apr. 2010.
164. M. Lucente and T. A. Galyean. Rendering interactive holographic images. In *Proceedings of the 22nd Annual Conference on Computer Graphics and Interactive Techniques*, pages 387–394. ACM, Sep. 1995.
165. C. Petz and M. Magnor. Fast hologram synthesis for 3d geometry models using graphics hardware. In *Proc. SPIE*, volume 5005, pages 266–275. International Society for Optics and Photonics, Jan. 2003.
166. V. M. Bove Jr, W. J. Plesniak, T. Quentmeyer, and J. Barabas. Real-time holographic video images with commodity pc hardware. In *Proc. SPIE*, volume 5664, pages 255–262. International Society for Optics and Photonics, Jun. 2005.
167. N. Masuda, T. Ito, T. Tanaka, A. Shiraki, and T. Sugie. Computer generated holography using a graphics processing unit. *Opt. Express*, 14(2):603–608, Jan. 2006.
168. T. Shimobaba, Y. Sato, J. Miura, M. Takenouchi, and T. Ito. Real-time digital holographic microscopy using the graphic processing unit. *Opt. Express*, 16(16):11776–11781, Aug. 2008.
169. L. Ahrenberg, A. J. Page, B. M. Hennelly, J. B. McDonald, and T. J. Naughton. Using commodity graphics hardware for real-time digital holo-

gram view-reconstruction. *J. Disp. Tech.*, 5(4):111–119, Apr. 2009.

170. C. Oh, S. O. Isikman, B. Khademhosseinieh, and A. Ozcan. On-chip differential interference contrast microscopy using lensless digital holography. *Opt. Express*, 18(5):4717–4726, Mar. 2010.
171. N. Takada, T. Shimobaba, H. Nakayama, A. Shiraki, N. Okada, M. Oikawa, N. Masuda, and T. Ito. Fast high-resolution computer-generated hologram computation using multiple graphics processing unit cluster system. *Appl. Opt.*, 51(30):7303–7307, Oct. 2012.
172. K. Onda and F. Arai. Multi-beam bilateral teleoperation of holographic optical tweezers. *Opt. Express*, 20(4):3633–3641, Feb. 2012.
173. T. Shimobaba, J. Weng, T. Sakurai, N. Okada, T. Nishitsuji, N. Takada, A. Shiraki, N. Masuda, and T. Ito. Computational wave optics library for C++: CWO++ library. *Comput. Phys. Commun.*, 183(5):1124–1138, May. 2012.
174. K. Murano, T. Shimobaba, A. Sugiyama, N. Takada, T. Kakue, M. Oikawa, and T. Ito. Fast computation of computer-generated hologram using xeon phi coprocessor. *Comput. Phys. Commun.*, 185(10):2742–2757, Oct. 2014.
175. A. Sugiyama, N. Masuda, M. Oikawa, N. Okada, T. Kakue, T. Shimobaba, and T. Ito. Acceleration of computer-generated hologram by greatly reduced array of processor element with data reduction. *Opt. Eng.*, 53(11):113104–113104, Nov. 2014.

Index